International Association of Hydrogeologists



**Jaroslav Vrba**
**Alexander Zoporozec**
**(Editors)**

# Guidebook on Mapping Groundwater Vulnerability

Volume 16 / 1994
International Contributions to Hydrogeology
Founded by
G. Castany, E. Groba, E. Romijn

Verlag Heinz Heise

Die Deutsche Bibliothek – CIP-Einheitsaufnahme

**Guidebook on mapping groundwater vulnerability** /
International Association of Hydrogeologists. Jaroslav Vrba;
Alexander Zaporozec, ed. – Hannover : Heise, 1994
(International contributions to hydrogeology ; Vol. 16)
ISBN 3–922705–97–9
NE: Vrba, Jaroslav [Hrsg.]; International Association of
Hydrogeologists; GT

**Financially supported by UNESCO as a contribution to the International Hydrological Programme IHP-IV, Project M-1.2 (a)**



This "Guidebook on Mapping Groundwater Vulnerability"
was prepared by members of the IAH/IHP Joint Working Group
as part of IHP-IV Profject M-1.2 (a)

J. Vrba, Czech Republic, Chairman
B. Adams, United Kingdom
R. Aldwell, Ireland
M. Civita, Italy
W. H. Gilbrich, Unesco, France
E. Gosk, Denmark
J. Hahn, German Federal Republic
J. Khouri, Syrian Arab Republic
G. Matthess, German Federal Republic
J. C. Miller, United States
H. G. van Waegeningh, The Netherlands
A. Zaporozec, United States

Volume 16, 1994
International Contributions to Hydrogeology
Founded by G. Castany, E. Groba, E. Romijn
ISSN 0936-3912
ISBN 3-922705-97-9
Printed by R. van Acken GmbH, Josefstraße 35, D-49809 Lingen

P.O.B. 610407, D-30604 Hannover

**This book is dedicated to the memory of**

**Lars Jørgen Andersen**

**-- a devoted, longtime member of the International Association of Hydrogeologists and its Commission for Ground Water Protection -- who died before this book was completed**

## MEMBERS OF THE IAH/IHP JOINT WORKING GROUP ON MAPPING GROUNDWATER VULNERABILITY


Brian Adams
British Geological Survey
Maclean Building
Wallingford, Oxfordshire
0X10 8BB United Kingdom

Robert Aldwell
Geological Survey of Ireland
Beggars Bush, Haddington Road
Dublin 4, Ireland

Massimo Civita
Dipartmento di Georisorse e Territorio
Politecnico di Torino
24 Corso Duca degli Abruzzi
10129 Torino, Italy

Wilfried H. Gilbrich
Division of Water Sciences
UNESCO
Place de Fontenoy
75700 Paris, France

Edmund Gosk
Geological Survey of Denmark
8 Thoravej
DK-2400 Copenhagen NV, Denmark

Jürgen Hahn
Bundesanstalt für Geowissenschaften und Rohstoffe
Postfach 510153
D-30655 Hannover, Germany

Jean Khouri
Arab Center for the Studies of Arid Zones and Dry Lands
PO Box 2440
Damascus, Syrian Arab Republic

Georg Matthess
Geologisch-Palaeontologisches Institut
Universität Kiel
Olshausenstr. 40/60
D-24098 Kiel, Germany

John C. Miller
CH2M HILL, Inc.
2701 N. Rocky Point Drive, Suite 800
Tampa, FL 33607, USA

Jaroslav Vrba
Korandova 32
14700 Praha 4
Czech Republic

Hubert G. van Waegeningh
Lange Voort 46
2341 KB Oegstgeest
The Netherlands

Alexander Zaporozec
University of Wisconsin-Extension
Wisconsin Geological & Natural History Survey
3817 Mineral Point Road
Madison, WI 53705, USA

## CONTENTS

FIGURES

TABLES

## FOREWORD

A picture speaks more than a thousand words and a map more than a thousand pictures. Maps, therefore have become an indispensable tool for water scientists and professionals to document data and describe hydrological situations. The art of hydrological mapping has developed historically from geological mapping. During the 1960s and 1970s a great deal of progress was made in the field of hydrogeological mapping. Efforts were made to develop a methodology, to compile model legends, and to print small-scale maps. UNESCO, IAH, and IAHS joined forces and a number of continental, regional, and national maps resulted from their cooperation. Particular milestones were the 1977 UNESCO/WMO publication *Hydrological Maps* and, in 1970, the *International Legend for Hydrogeological Maps* issued by UNESCO, IAHS, IAH, and the British Institute of Geological Sciences in 1970, and slightly modified in 1983.

As a result of the rapid development in hydrogeological mapping and the numerous applications in very different hydrogeological regions, the above publications have become obsolete. As a contribution to the International Hydrogeological Programme (IHP), the International Association of Hydrogeologists (IAH) has compiled a completely new guidebook on hydrogeological mapping with a revised model legend. This book is supposed to become a new standard work and will be published by IAH as Volume 17 of the International Contributions to Hydrogeology. Another guidebook, on mapping of surface water resources, is now being prepared by a special IHP Working Group.

While the above publications deal with hydrological mapping in general, a need has also arisen to prepare guidelines for special applications. Aquifers are no longer natural water bodies. The ever growing demand for water today means that not even the smallest aquifer is forgotten -- intensive exploitation has become the rule. Protection has become essential, particularly with regard to aquifer contamination brought about by human interference. Some aquifers are well protected by nature through dense covering layers, others are extremely vulnerable.

At the Budapest meeting of the IAH Ground Water Protection Commission in 1987, an idea was introduced to include among the future Commission activities the topic of groundwater vulnerability assessment and mapping. The initial position paper was prepared in 1988 by H.G. van Waegeningh. The paper and the topic of vulnerability mapping was discussed at the Commission meeting in Czechoslovakia in 1989, where a Project Working Group was established and a tentative content of a guidebook was developed. At the next Commission meeting in The Netherlands in 1990 an outline of a report on groundwater vulnerability maps was prepared and authors of chapters were tentatively selected.

At the same time, UNESCO prepared the fourth phase (1990-1995) of the International Hydrological Programme (IHP). Within subprogramme M "Management of Water Resources for Sustainable Development", Theme 1 is related to methodologies for water resource assessment and hydrological design. One of the projects under this theme, namely

Project M-1.2(a), foresees the preparation of a methodological guide for mapping groundwater resources and their vulnerability. In order to avoid duplication of efforts and parallel work, UNESCO suggested merging the IAH and IHP working groups and to jointly prepare the guidebook. A fee contract between UNESCO and IAH was signed in 1991.

Because of the inclusion of the Commission project on vulnerability maps into IHP-IV Project M-1-2a, it was necessary to modify its original schedule and objectives to make it fit to UNESCO IHP goals. The first meeting of the IAH/IHP Joint Working Group was held in Tampa, Florida, USA in April 1991. The outline and schedule of the study were revised, the individual chapters reassigned to authors, and Dr. Vrba and Dr. Zaporozec were appointed as editors of the report. First drafts of the chapters were reviewed at the 1992 meeting of the joint group in Torino, Italy. A subgroup met at Leiden, The Netherlands in early 1993, in order to prepare the model legend. The revised chapters were edited during 1993 and approved at the editorial group meeting in Norway in June 1993 and at the joint group meeting in Wallingford, UK in May 1994.

Besides the listed authors, many other members of the Ground Water Protection Commission of IAH provided valuable suggestions during the preparation of this study and participated in the review of the final report. Stimulating ideas and recommendations concerning its contents were presented at discussions held during the Commission sessions in 1989 (Skály, Czechoslovakia), 1990 (Bilthoven, The Netherlands); and 1991 (Tampa, Florida, USA). Thanks are expressed to UNESCO for funding the project; to the Wisconsin Geological and Natural History Survey for the administrative and technical support; and particularly, to Mr. W.H. Gilbrich, Project Officer, UNESCO Division of Water Sciences, who cooperated actively in the realization of the IAH/UNESCO Project "Preparation and Use of Ground Water Vulnerability Maps".

*Jaroslav Vrba*
*Chairman*
*IAH Commission for Ground Water Protection*

## LIST OF ACRONYMS AND ABBREVIATIONS

| | |
|---|---|
| AM/FM | Automated Mapping/Facilities Management, a computer-assisted system of mapping, generally employed in the depiction and analysis of relationships between networked fractures, primarily in a linear form. |
| Amer. | American. |
| Assoc. | Association. |
| B.R.G.M. | Bureau de Recherches Géologiques et Minières (French Agency for Geological and Mining Survey). |
| Bull. | Bulletin. |
| CAD | Computer Aided Design, a computer-assisted system of designing and mapping, not capable of managing or analyzing the descriptive information associated with those features drawn with it. |
| Conf. | Conference. |
| DBMS | Data Base Management System, main software for row data organizing and management of GIS. |
| DRASTIC | A standardized, rating system evaluating groundwater contamination potential of selected hydrogeological settings based on seven factors: Depth to Water, Net Recharge, Aquifer Media, Soil Media, Topography, Impact of Vadose (Unsaturated) Zone, and Hydraulic Conductivity of the Aquifer (Aller et al, 1987). |
| ed(s). | Editor(s). |
| EDP | Electronic Data Processor. |
| E(E)C | European (Economic) Community (now called European Union). |
| e.g. | for example (Latin *exempli gratia*). |
| et al. | and others (Latin *et alii*). |
| Env. | Environmental. |
| etc. | and so forth (Latin *et cetera*). |

Ges. Gesellschaft.

GIS Geographic Information System, a computer system of geographically organized, polygonal spatial data for interactive processing, storage, and real time mapping.

G.N.D.C.I.-C.N.R. Gruppo Nazionale per la Difesa dalle Catastrofi Idrogeologiche-Consiglio Nazionale delle Richerche (Italian National Group for the Prevention of Hydrogeological Disasters-National Research Council).

GOD An empirical ranking system for the rapid assessment of aquifer vulnerability to contamination, developed by Foster (1987) and based on three factors: Groundwater Occurrence (type of aquifer), Overall Aquifer Class (grade of consolidation and lithological character), and Depth to Groundwater.

HCMM Heat Capacity Mapping Mission satellite boarding a two-channel scanning radiometer (near and thermal infrared).

IAEA International Atomic Energy Agency.

IAH International Association of Hydrogeologists.

IAHS International Association of Hydrological Sciences.

i.e. that is (Latin *id est*).

I.G.M.E. Instituto Geologico y Minero de Espagna (Geological and Mining Survey of Spain).

IHP International Hydrological Programme.

Inst. Institute.

Intl. International.

Jour. Journal.

MSS Multispectral Scanner, a line scanner that simultaneously records image data in several wavebands carried by the LANDSAT series of satellites.

n. (no.) Number.

p. Page.

PCSM Point Count System Model, a parametric system based on rating and weighting of selected parameters to assess aquifer vulnerability to contamination.

Proc. Proceedings.

PSC Pollution Source Center, a human-made activity, point or/and nonpoint, existing or potential source of groundwater contamination.

Publ. Publication.

R.I.V.M. Rijksinstituut voor Volksgezondheid en Milieuhygiene (National Institute of Public Health and Environmental Hygiene of The Netherlands).

SAR Synthetic-Aperture Radar.

Sci. Science.

SINTACS A computer-assisted, point count system model for the assessment of aquifer vulnerability to contamination developed in Italy and based on seven factors: Soggiacenza (depth to water table), Infiltrazione (infiltration), Azione del Non Saturo (unsaturated zone function), Tipologia della Copertura (soil cover), Caratteri Idrogeologici dell' Acquifero (hydrogeological characteristics of the aquifer), Conducibilità Idraulica (hydraulic conductivity), and Acclività della Superficie Topografica (average slope of the topographic surface).

SIR Shuttle Imaging Radar, SAR experiments carried aboard the NASA Space Shuttle.

SMIRR Shuttle Multispectral Infrared Radiometer, a non-imaging spectroradiometer carried by the NASA Space Shuttle and covering wavebands in the 0.5-2.4 $\mu$m range.

Soc. Society.

SPOT Satallite Probatoire pour l'Observation de la Terre, a French satellite carrying two imaging pointable systems allowing stereoscopic viewing of the Earth.

Symp. Symposium.

TIN Triangulated Irregular Network, subsystem of a GIS to process row data in contoured map format.

Unesco United Nations Educational, Scientific and Cultural Organization.

U.S. United States (of America).

U.S.E.P.A. United States Environmental Protection Agency.

| | |
|---|---|
| v. (vol.) | Volume. |
| VOC | Volatile organic compound. |
| WMO | World Meteorological Organization. |
| Z. | Zeitschrift. |

## EXECUTIVE SUMMARY
## GUIDEBOOK ON MAPPING GROUNDWATER VULNERABILITY

### Introduction

The Ground Water Protection Commission of the International Association of Hydrogeologists (IAH) in 1987 introduced among future Commission activities a project that lead to the preparation and publication of this book. In the fourth phase (1990-1995) of the International Hydrological Programme (IHP), Unesco initiated a project on the preparation of a methodological guide for mapping groundwater resources and their vulnerability. In order to avoid duplication of efforts and parallel work, Unesco suggested merging the IHP and IAH projects. This book is the result of this cooperative effort.

This guidebook on groundwater vulnerability assessment and mapping intends to help primarily map makers in designing and compiling vulnerability maps and to help users of the maps understand their contents and value. The methodology of vulnerability assessment and mapping presented in the book attempts to provide them with a comprehensive guide to interpretation of hydrogeological and other relevant data and with an understandable format of presenting the data. In order to facilitate the preparation of consistent, uniform, and comparable vulnerability maps, a model legend for groundwater vulnerability maps is also included in this book.

The book is accompanied by examples of vulnerability assessment and maps; an extensive list of references; a glossary to help explain some less common technical terms; and a list of acronyms and abbreviations.

### Concept of Groundwater Vulnerability

The concept of groundwater vulnerability is based on the assumption that the physical environment may provide some degree of protection to groundwater against the natural and human impacts, especially with regard to contaminants entering the subsurface environment. The term "vulnerability of groundwater to contamination" was introduced by French hydrogeologist J. Margat in the late 1960s. The idea of describing the degree of vulnerability of ground water to contaminants as a function of hydrogeological conditions by means of maps was conceived to show that the protection provided by the natural environment varies at different locations.

The fundamental concept of groundwater vulnerability is that some land areas are more vulnerable to groundwater contamination than others. The ultimate goal of the vulnerability map is a subdivision of an area into several units showing the differential potential for a specified purpose and use. Results of vulnerability assessment are portrayed on a map showing various homogeneous areas, sometimes called cells or polygon, which have different levels of vulnerability. The differentiation between the cells is, however, arbitrary because vulnerability maps only show relative vulnerability of certain areas to others, and do not represent absolute values.

Although the concept of groundwater vulnerability has been around for almost three decades, a generally recognized and accepted definition of vulnerability has not been developed yet. Attempts to define vulnerability started in the 1980s, and are summarized and analyzed in this book. For this book, the following definition is used: "Vulnerability is an intrinsic property of a groundwater system that depends on the sensitivity of that system to human and /or natural impacts."

There may be more than one type of groundwater vulnerability. In this book the term "intrinsic (or natural) vulnerability" is defined solely as a function of hydrogeological factors--the characteristics of an aquifer and of the overlying soil and geological materials. In addition to intrinsic properties of a groundwater system, some users of vulnerability maps may also wish to include potential human impacts, which may prove detrimental--in space and time--to the present and future uses of the groundwater resource. For this concept, the term "specific (or integrated) vulnerability is used."

**Chemical and Biological Contaminants and Their Subsurface Behavior**

Description of groundwater vulnerability to contamination requires evaluation of the potential for contaminant attenuation. The earth materials have certain capacity to remove some contaminants or reduce their concentration. This "purification capacity" of the environment is called "attenuation capacity" and it expresses the intrinsic ability of the earth materials above and within the groundwater system to adsorb, disperse, or retard contaminants by a number of physical, chemical, and biological processes acting in the soil-rock-groundwater system.

There are several groups of potential contaminants each affected by different attenuation processes. Certain contaminants appear to predominate in groundwater. They are heavy metals, nutrients, organic chemicals, fertilizer and pesticide constituents, salt, bacteria, and viruses.

The attenuation of contaminants as they travel through the soil zone, unsaturated zone, and groundwater zone is affected by a variety of naturally occurring chemical reactions and biological and physical processes that often cause the contaminant to change its physical state or chemical form. The principal reactions and processes include geochemical processes (such as adsorption/desorption, solution/precipitation, and oxidation/reduction); physical processes (such as advection, dispersion, retardation, and filtration); biochemical processes (organic decomposition and cell synthesis); and biophysical processes (filtration and transport of pathogens).

The occurrence and intensity of these processes vary in the subsurface. The soil zone has the greatest variety and magnitude of natural processes, especially in the root zone where significant amounts of chemicals are broken down by microorganisms or by chemical and physical processes and taken up by plants. The unsaturated zone usually plays most important role in delaying the arrival of contaminants to the water table. Fewer processes take place in the saturated zone where solution, dilution, and hydrodynamic dispersion are most effective.

## Classification and Review of Groundwater Vulnerability Maps

Groundwater vulnerability maps belong to the category of special-purpose environmental maps. They are classified as interpretive groundwater protection maps, derived from general hydrogeological maps. They differ from hydrogeological maps in that they do not show the elements of a groundwater system but the specific characteristics of these elements as they relate to vulnerability of ground water. The ultimate goal of a vulnerability map is a subdivision of an area into several units showing the differential potential for a specified purpose and use. Groundwater vulnerability maps are time dependent, requiring updating to portray changes in a groundwater system and in the location and nature of human impacts.

Purpose, scale, contents, and methods of graphical representation are the most important criteria for the classification of vulnerability maps. The scale of vulnerability maps controls the map contents and can be selected according to the purpose of maps, the character and complexity of hydrogeological conditions, and the accuracy required for problem solving.

The first concepts and methodology of groundwater vulnerability maps were developed in the mid-1960s in Europe. In the late 1960s and in the 1970s, the compilation of vulnerability maps were introduced in several European countries (primarily in France, Czechoslovakia, and Germany) and the USA. These first maps were mostly of small scale, covering the entire state territory or large regions.

The later phase in the development of vulnerability maps was characterized by the transition to vulnerability maps on medium and larger scale, covering smaller territorial units. Since the early 1980s a considerable number of groundwater vulnerability maps have been published throughout the world, expressing mostly groundwater vulnerability to contamination.

## Assessment of Groundwater Vulnerability

Vulnerability of groundwater is a relative, non-measurable, dimensionless property. The accuracy of vulnerability assessment depends above all on the amount and quality of representative and reliable data.

The principal attributes used in the assessment of intrinsic groundwater vulnerability are recharge, soil properties, and characteristics of the unsaturated and saturated zone. Attributes of secondary importance include topography, ground water/surface water relation, and the nature of the underlying unit of the aquifer.

Specific vulnerability is mostly assessed in terms of the danger of the groundwater system becoming exposed to contaminant loading. The contaminant's travel time in the unsaturated zone and its residence time in the aquifer, and the attenuation capacity of the soil-rock-ground water system with respect to the properties of individual contaminants, are the most important parameters in the assessment of specific groundwater vulnerability.

Methods and techniques available for the assessment of groundwater vulnerability vary according to the physiography of the study area, purpose of the study, and quantity and quality of data. The available methods can be grouped into three basic categories:

(a) The hydrogeological setting methods are an universal system suitable for large areas with a variety of natural features. The methods involve the comparison of a subject area to criteria judged to represent vulnerable conditions.

(b) The parametric methods include matrix systems, rating systems, and point count systems. The overall procedure for all of these systems begins with the selection of parameters judged to be representative for vulnerability assessment. A multiplier (importance weight) may be assigned to each parameter to reflect the relationships among the parameters and their importance for vulnerability assessment. Each of the selected parameters has a given range, which is subdivided into discrete hierarchical intervals. Each interval is assigned a value reflecting the relative degree of vulnerability, and the rating points are summed. The final numerical score is divided into segments expressing a relative vulnerability degree.

(c) Analogical relations and numerical models are based on mathematical symbols resulting in a vulnerability index. These methods are generally applicable for the assessment of specific vulnerability only.

Vulnerability assessment should be based on hydrogeological evaluation rather than on general, automatic rating procedures. A combination of aquifer simulation models and geographical information systems offer excellent advantage to perform these tasks.

**Groundwater Vulnerability in Areas of Climatic Extremes**

Extreme dryness, extreme wetness, extreme heat, and extreme cold are the principal conditions that influence groundwater vulnerability in regions of climatic extremes.

Groundwater vulnerability is particularly important in arid zones, which are very sensitive to even minor shifts in their water balance. Groundwater in these zones is particularly vulnerable to drought. Recharge, and its amount and mode, is the attribute of primary importance when vulnerability of groundwater to drought needs to be assessed. Areas of aquifers that are sensitive to drought are usually also the most vulnerable to desertification. The assessment of groundwater vulnerability to contamination, salinization, and depletion could assist planners in minimizing adverse effects of ground water development in arid zones.

Regions of extreme wetness receive a large quantity of rainfall and have potential for greater recharge. Groundwater vulnerability in such regions is high because contaminants tend to quickly enter into solution and rapidly move downward to the water table. At the same time, there is a lesser soil-rock contact time for contaminants.

In regions of extreme heat (either dry or wet), solubility plays a major role in vulnerability of groundwater because it generally increases with temperature. The contaminants and groundwater will be warmer and may be more reactive with the soil-rock materials.

In regions of extreme cold, frozen soils tend to inhibit the downward movement of contaminants and reduce groundwater vulnerability. When contamination occurs in permafrost regions, "self-cleaning" ability of groundwater is low and contaminants tend to concentrate rather than becoming diluted.

**Data Needs and Presentation**

Vulnerability assessment requires a thorough knowledge of hydrogeological and hydrochemical data and location of potential contamination sources. In many cases data may be available from government agencies, universities, research institutes, state or provincial geological surveys, and resource exploration and consulting companies. However, in some areas the representative data are not available, and therefore, they have to be acquired through field measurements and observations.

The amount, quality, and distribution of basic data determines the quality and accuracy of vulnerability assessment. A close correlation exists between the density of data points, amount of data obtained at any measured point, and the scale at which the map is to be constructed. For example, an assessment and mapping of small areas requires a large- scale map with high density of data points per unit area. On the other hand, for areas where the density of data points is low, only simple assessment methods can be used, resulting into small-scale maps.

Also the reliability of basic data has to be considered before selecting a method for vulnerability assessment. For example, reliability of data varies with the elevation of the area under study, and it sharply decreases already at altitudes higher then 300 m above the sea level. Therefore, for mountainous regions, only the simpler assessment can be used.

Vulnerability maps can be created manually or photographically, if the individual data layers are on transparencies; or by a computer, if the data are encoded into a geographical information system (GIS).

An important step in manual vulnerability mapping is the method of combining data that are being mapped. One of the most widely used approaches is the overlay method, which involves producing several maps of individual attributes or their parameters on scale-stable transparent material. A composite vulnerability map can be obtained by stacking all of the transparencies.

Stacking of the individual layers of data also can be done by computer. Data can be manipulated by any of the existing GIS, such as ARC/INFO, ERDAS, or GENE MAP. This approach requires that all attributes and their parameters are geographically referenced, digitized, and entered into a data base. Once in the data base, it is possible to register all data sets as data layers with a common coordinate system and manipulate them to produce derivative maps, and finally, a vulnerability map.

The design of groundwater vulnerability maps still lacks international coordination and standardization. An agreement should be reached on colors, patterns, and symbols to be used; a standard graphical design; and explanatory notes and text accompanying the map. Explanatory notes and text are an integral part of a vulnerability map. All information should be presented on one map sheet. The map sheet should consist of the main vulnerability map with a short and long legend. The short, synoptical legend summarizes in several columns the individual degrees of vulnerability portrayed on the map. The longer legend, using the same colors and patterns included in the short legend, explains the vulnerability categories in a greater detail.

The map sheet also may include the explanation of symbols of human activities affecting the groundwater; cross sections or block diagrams; and a large-scale maps showing specific information, such as existing contamination conditions, groundwater quality, or land-use patterns; or small-scale maps of individual attributes used in the preparation of the vulnerability map.

**Uses and Limitations of Groundwater Vulnerability Maps**

Groundwater vulnerability maps are particularly useful for regulatory, managerial, and decision-making purposes at all level of government. First of all, vulnerability maps can help planners and regulators make informed, environmentally sound decisions regarding land use and groundwater protection. The vulnerability maps are frequently combined with land-use maps, groundwater quality data, and contamination source inventories to direct available financial and manpower resources to the most vulnerable areas.

Vulnerability maps are a good tool to make local and regional assessment of groundwater vulnerability potential, to identify areas susceptible to contamination, to design monitoring networks, and to evaluate ground water contamination, particularly nonpoint contamination.

Vulnerability maps also are helpful for educating and informing planners, managers, and decision- and policy-makers about groundwater protection, risk of contamination, and contamination prevention. The maps also can be used to educate the public about ground water being part of a larger, interconnected ecological system.

The limitations of vulnerability maps are mainly caused by the lack of representative data (both in terms of amount and quality) and their relation to the scale at which the map has been compiled; inadequate description of the physical (particularly geological and hydrogeological) system; lack of generally accepted methodology; and limited verification and control of vulnerability assessment methods due to the long time involved in the processes affecting groundwater vulnerability.

Vulnerability maps should be carefully thought out and their meaning and degree of reliability fully explained. It is important that disclaimers appear on maps informing the user of the map limitations and intended use. The map also should be accompanied by description of the assumptions and methodologies used and the level of accuracy of presented information. With a proper disclaimer, any vulnerability map can be used, even that one based on scanty data. However, under no circumstances should the vulnerability maps be used as substitutes for site-specific studies.

In order to have a broad range of uses and applications, vulnerability maps should be consistent, standardized in graphical and numerical expression, understandable, with a good legend and comprehensive explanatory notes, thereby helping overcome the gap that frequently exists between the scientific and lay communities. Vulnerability maps are compiled for practical uses, therefore, they cannot be too sophisticated and overcrowded with data, which may lead to their misinterpretation or misuse.

**Future Trends in Groundwater Vulnerability Mapping**

Before considering possible future trends in groundwater vulnerability mapping, there are a number of underlying issues that remain to be resolved. Three of the most important are:

1) Development of a generally recognized and accepted definition of vulnerability. Presently, there are many terms describing groundwater vulnerability, which create confusion as to the real meaning of groundwater vulnerability.

2) Agreement on a generally acceptable approach to vulnerability mapping and consistency in the use of methods and symbols expressing vulnerability on maps. The methodologies for the preparation of groundwater vulnerability maps are still being developed and more work and international cooperation is needed to determine an uniform approach to interpretation and assessment of a standard set of mappable attributes; optimum map contents; and a map format. The use of common sets of vulnerability maps would improve the consistency and comparability across similar studies and similar areas.

3) Testing the validity of vulnerability maps. To date little has been done to verify how accurately the existing vulnerability maps correspond to actual situation. Careful field monitoring will be needed to test predictions and thus enable further refinement of the assessment and mapping concepts.

The availability of computers now enables the easy and rapid handling of large amounts of data. Furthermore, the new digital mapping technology has revolutionized the manipulation of data. These developments will lead to the:

- improvement of vulnerability assessment methods;
- standardization of methods to obtain basic attributes;
- greater quantitative precision in defining vulnerability classes based on the knowledge of groundwater flow and contaminant transport;
- increased production of large-scale, specific vulnerability maps;

- regular and rapid updating of existing maps as new information becomes available;
- integration of vulnerability maps on a routine basis into local and regional planning procedures and decision-making.

**Model Legend for Groundwater Vulnerability Maps**

A model legend was proposed to facilitate the preparation of groundwater vulnerability maps in an internationally standardized form. In order to develop a groundwater vulnerability map, it is useful to categorize the basic information relating to vulnerability into primary and secondary.

The primary information relates to the intrinsic vulnerability of the groundwater based on the nature and thickness of the strata overlying an aquifer and is represented on the map by a full-color shading. Five classes of vulnerability are recommended: extremely high, high, medium, low, and very low, represented by the red orange, rose, yellow, light green, and dark green colors, respectively. Careful choice of colors is required in order to permit legibility of superimposed patterns, ornaments, and symbols. Bright colors are optically appealing, but experience has shown that the use of less intense colors is generally more effective.

In addition, a soil classification system can be included by employing different tones of the colors. The soil classification, when required, needs only be superimposed upon the extremely high, high, and medium classes of vulnerability. A non-aquifer is represented by brown shading, which overrides any consideration of the unsaturated zone.

The secondary information relates to the potential for contamination and is based on a consideration of the nature of the saturated zone. This information is superimposed as an ornament (pattern) on the basic shading representing the vulnerability class.

A series of symbols is recommended for other relevant data, such as hydrogeological features, water-supply objects needing protection, potentially contaminating human activities, and the existing quality status of groundwater.

The proposed ornaments presented in the model legend are generally applicable to medium-scale or large-scale maps (between 1:200 000 and 1:25 000). Maps can be accompanied by diagrams, cross sections, and side maps. It is strongly recommended that the map, legend, and explanatory notes form an inseparable unit, i.e. be printed on one sheet.

**Examples of Groundwater Vulnerability Assessment and Maps**

In order to illustrate the application of vulnerability assessment and its potential misinterpretation, five examples of vulnerability assessment in various aquifer conditions and stresses are presented. In each example the same assessment procedure was followed. Some of the examples show that, in spite of the many positive applications of groundwater vulnerability assessment, there is a danger of faulty management decisions if these decisions are based solely on ready-made, generalized vulnerability maps.

A portion of the groundwater vulnerability map of East Kent, UK, scale 1:100 000, is included as an example of a medium-scale, operational type of map. It includes details of the soil and geological classifications and the way they are combined to give the vulnerability classification.

Chapter 1.
# INTRODUCTION

The quality of groundwater is receiving widespread attention all over the world; and hydrogeological information is essential to the effective protection and management of groundwater quality. Effective protection should be aimed at the prevention of problems and requires a sound information base to determine, on a continuing basis, the groundwater quality problems that exist and those that may develop in the future. Groundwater vulnerability maps are important tools to assist in relaying this information. It is important to remember that vulnerability maps are not "panacea"; they are simply just one of the many tools available for groundwater protection programs.

Groundwater vulnerability maps belong to the group of groundwater protection maps, which are one of the most important categories of special-purpose environmental maps. They are derived from general hydrogeological maps but differ from these maps in that they are interpretive and user-oriented. Groundwater vulnerability maps are graphical interpretations of the natural attributes of groundwater systems for specific areas and specific purposes. The principal natural attributes include properties of the soil, unsaturated zone, and aquifer and amount of recharge to groundwater.

The concept of groundwater vulnerability maps relies on the assessment and representation of these attributes and depends on a given scenario and objectives for which a particular map is being compiled. The formulation and definition of what we understand by groundwater vulnerability is of essential importance for the map concept and design, selection of methods of data presentation, and map compilation. Aquifer vulnerability is usually assessed, and most groundwater vulnerability maps compiled, with regard to contamination resulting from human activities. However, a groundwater system may also be vulnerable to climatic and other natural processes.

In the objectives of IHP-IV Project M-1.2(a), the concept of vulnerability is introduced both in terms of quality (contamination) and quantity (water depletion). In fact, it is often very difficult to separate the qualitative and quantitative aspects of groundwater vulnerability. For example, overexploitation of an aquifer system need not become expressed only in the quantitative terms (decline of the water table or change in the groundwater flow system) but also in a changed composition of groundwater (a qualitative aspect). Cartographical representation of groundwater vulnerability, in the view of the objectives of the IHP/UNESCO project, should therefore, include vulnerability to human impacts as well as to natural processes and the quantitative and qualitative aspects of vulnerability. Nevertheless, it is clear that worldwide concern in years to come will be more focused on the potential contamination of groundwater. Therefore, for practical reasons, this report deals mainly with the aspects of vulnerability related to the quality of groundwater.

Vulnerability maps are a valuable planning tool to overcome problems of haphazard, uncontrolled development of land and of undesirable activities having an impact on groundwater quality. They support planning, regulatory, managerial, and decision-making activities and are of great value to environmental specialists, consultants, engineers, and hydrogeologists responsible for solving problems related to groundwater management and protection. However, vulnerability maps can only give a general view, and not a specific detail that planners, managers, or local officials seek for solution of their problems; and misinterpretation of maps can give the planners and administrators a false sense of security. In order to reduce their misinterpretation and misuse to a minimum, vulnerability maps must carry a warning about their limitations and include instructions on how to use them.

IAH and UNESCO decided to prepare a guidebook on groundwater vulnerability mapping to help primarily map makers in designing and compiling vulnerability maps and to help users of vulnerability maps understand map contents and value. The methodology of vulnerability mapping presented in this book attempts to provide them with a comprehensive guide to interpretation of hydrogeological and other relevant data and with an understandable format of presenting the data.

By its very nature a guidebook is narrative. The primary purpose of this book, however, is to advise map makers; and the tool to translate hydrogeological facts into the language of a map is its legend. The authors, therefore, have developed a model legend keeping in mind that (a) the legend must be clear, concise, and comprehensive; (b) the symbols must express vulnerability values using cartographical analogies; and (c) the legend must conform to the *International Legend for Hydrogeological Maps* (Unesco, 1970) where applicable. The proposed model legend (Appendix A) is based on the experience gained in several countries. An attempt has been made to compile a legend universally applicable.

All co-authors have long-standing experience in applied hydrogeology and in questions related to aquifer protection. This explains why the guide has been written by practitioners for practitioners. The authors are, however, aware that this book represents a first attempt to describe a complicated matter. They would be grateful for whatever suggestions the reader may wish to offer for its improvement.

The authors believe that there is no reason to write a methodological handbook on vulnerability mapping at this point. Each situation is specific and requires a specific approach; a hydrogeologist in charge of the project must decide which type and scale of a map would fit the purpose best. Moreover, mapping techniques are well known, and it is not necessary to repeat them in this book. However, standardization, especially of the format and scale of maps and of the legend, and consistent approach to vulnerability mapping is desirable. The use of common methodology to produce vulnerability maps would improve the consistency and comparability across similar studies as well as the efficiency of developing new maps for similar projects. The authors hope that with proper education and information, to which this book is aimed to contribute, vulnerability maps may become a useful and appropriate tool in the field of environmental protection and management.

Chapter 2.

# CONCEPT OF GROUNDWATER VULNERABILITY

Impetus for development of the concept of groundwater vulnerability has been generated by the emerging worldwide concern about the problems of groundwater contamination*. In the search for tools to deal with contamination of groundwater, the concept of groundwater vulnerability was introduced in the late 1960s. The original concept was called "vulnerability of groundwater to contamination" (Margat, 1968), and vulnerability has been usually assessed only with regard to contamination ever since. However, the concept of groundwater vulnerability is not related to contamination or water quality aspects only; it can also include water quantity aspects.

## NATURAL PROTECTION OF GROUNDWATER

The original concept of groundwater vulnerability was based on the assumption that the physical environment may provide some degree of protection to groundwater with regard to contaminants entering the subsurface. The earth materials may act as natural filters to screen out some contaminants. Water infiltrating at the land surface may be contaminated but is naturally purified to some degree as it percolates through the soil and other fine-grained materials in the unsaturated zone (Figure 1).

The degree of attenuation that occurs between the contaminant source and the aquifer determines the relative potential for groundwater contamination. The attenuation capacity, or "purification capacity", of subsurface materials consists of the interactions of numerous physical, chemical, and biological processes in a soil-rock-groundwater system and is

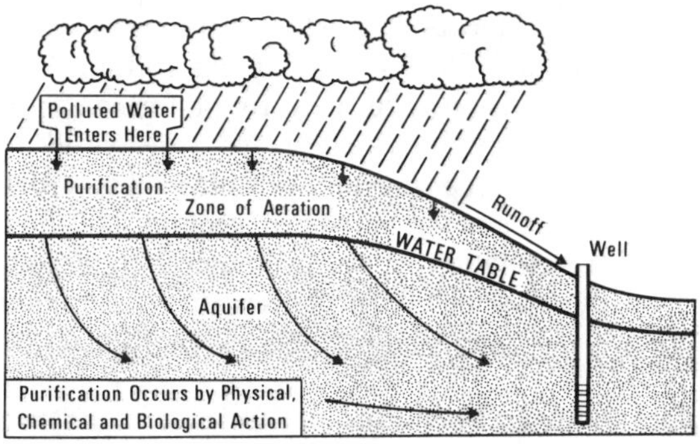


**Figure 1.** Natural purification of contaminated water.

-----------------------

*) For the sake of consistency, this report uses only the term "contamination" and not pollution. For the meaning of these terms, see Glossary.

significantly affected by the solute transport mechanism as well as hydrogeological conditions (Golwer, 1983). The principal natural processes affecting the transport and fate of solutes (or contaminants) in the subsurface are explained in Chapter 3.

The potential for natural protection is limited and extremely variable. Different parts of the physical environment have varying capacities for attenuating contaminants. Mapping the vulnerability or sensitivity of the physical environment enables us to identify areas that are more (and less) sensitive to contamination because of the materials overlying the groundwater.

Assessment of groundwater vulnerability to contamination has been widely used, and sometimes misused, as a tool to manage potential contamination of aquifers. However, the methods of assessment and its goals are often only implicitly stated (Pfannkuch, 1989). In order to treat the problem in the most rational and consistent way, it is necessary to formalize assessment procedures. Vulnerability assessment is generally done by mapping the contamination potential of the physical environment, with or without a rating system (Zaporozec, 1989), using several physical factors to evaluate the contamination potential. Recently more comprehensive assessment methods have been developed, based on risk analysis (Foster, 1987; Foster and Hirata, 1988; Pfannkuch, 1989; Trojan and Perry, 1988). Vulnerability assessment methods are discussed in Chapter 5.

## HISTORICAL OVERVIEW

The term "vulnerability of groundwater to contamination" was introduced in France in the 1960s (Margat, 1968). The idea of describing the degrees of vulnerability of groundwater to contaminants as a function of hydrogeological conditions by means of maps was conceived in an effort to create awareness of danger of groundwater contamination (Albinet and Margat, 1970; Margat, 1968). The aim of these maps was to show that the protection provided by the natural environment varied at different locations and to identify areas where protection measures were most needed (Margat and Suais-Parascandola, 1987). The visualization provided by maps proved to be an effective way of delivering information to a fairly large audience of decision-makers and administrators. The maps were based on fundamental hydrogeological factors: depth to water table, permeability of surficial deposits, connection between ground and surface water, and the average velocity of groundwater flow (Margat and Suais-Parascandola, 1987). The interpretation of hydrogeological conditions in terms of vulnerability was qualitative and did not include processes of contaminant migration from surface to the groundwater.

Although the concept of groundwater vulnerability has been around for more than two decades, a generally recognized and accepted definition of vulnerability has not been developed yet. One of the earlier definitions found in the literature is that of Albinet and Margat (1970) who offered that aquifer vulnerability is the possibility of percolation and diffusion of contaminants from the ground surface into natural water-table reservoirs, under natural conditions. Olmer and Řezáč (1974) suggested that the vulnerability of groundwater is "the degree of endangering, determined by natural conditions and independent on present source of pollution." In their view, vulnerability depends in the unsaturated zone on vertical permeability and in the aquifer on hydraulic gradient and velocity of the groundwater flow. In a 1981 lecture Vrána (1984a) defined aquifer vulnerability as the "complex of surface and subsurface natural conditions influencing the movement of a pollutant toward the aquifer."

Villumsen et al (1983) proposed that groundwater vulnerability is "the risk of chemical substances--used or disposed on or near the ground surface--to influence groundwater quality." According to these authors, groundwater vulnerability depends on a series of parameters, dynamic as well as static. They emphasized that the chemical composition of the groundwater may be used as an indicator of vulnerability and recommended that chemical analyses of groundwater be used for preliminary verification of the vulnerability maps.

Attempts to define groundwater vulnerability were made in several papers presented at the 1987 International Conference on Vulnerability of Soil and Groundwater to Pollutants (van Duijvenbooden and van Waegeningh, 1987). Some authors proposed direct definitions, some only indicated what the groundwater vulnerability should be based on. Bachmat and Collin (1987) defined groundwater vulnerability to contamination as "the sensitivity of its quality to anthropogenic activities which may prove detrimental to the present and/or intended usage-value of the resource." They suggested that, by definition, vulnerability should be expressed in terms of the change in concentration of a given substance per unit increment in a given human activity. Sotorníková and Vrba (1987) understand the vulnerability of a hydrogeological system as "the ability of this system to cope with external, both natural and anthropogenic, impacts which affect its state and character in time and space." Civita defined the degree of intrinsic vulnerability as a possibility of infiltration and percolation of contaminants through the unsaturated zone (Benacchio et al, 1988).

Several authors mentioned what parameters the groundwater vulnerability should include. In their approach to classify vulnerability of the groundwater resources in Germany, Vierhuff et al (1981) based the aquifer vulnerability on two main aspects: the degree of protection against contamination from the surface by the overlying strata and the potential for purification of contaminated water in the aquifer (Vierhuff, 1981). To assess the vulnerability, the authors used three attributes: type of aquifer, location of the aquifer in the hydrological cycle, and the characteristics of the unsaturated zone or confining layers.

Goosens and van Damme (1987) considered the static and dynamic factors to be of the same importance. For Klaučo (1987), the most important was the variability factor of the groundwater flow; for Friesel (1987), it was recharge. According to Johnston (1988) the vulnerability of an aquifer to contamination from the near-surface contamination sources is controlled by the groundwater flow system, the hydrogeological framework, and the climate.

Vrba (1991) introduced time scale in the definition of vulnerability. He suggested that "vulnerability on the human time scale is an unchanging natural intrinsic property of the unsaturated and saturated parts of a groundwater system and depends on the ability or inability of this system to cope with natural processes and human impacts."

The concept of groundwater vulnerability has gradually evolved from the mere assessment of hydrogeological characteristics to the assessment of the contamination risk placed upon aquifers by human activities.

Foster (1987) offered a definition based on groundwater contamination risk, which he considered as the interaction between the natural vulnerability of an aquifer and the contaminant loading that is, or will be, applied to the subsurface environment as a result of human activity. He used the term "aquifer pollution vulnerability" to represent the intrinsic characteristics that determine the sensitivity of various parts of an aquifer to being adversely affected by an imposed contaminant load.

The combination of aquifer susceptibility to potential contamination and the presence and characteristics of contaminants also appears in the latest definitions. Palmquist (1991) defined groundwater vulnerability as "a measure of the risk placed upon the ground waters by human activities and the presence of contaminants" and stated that "without the presence of contaminants, even the most susceptible ground water is not at risk, and thus, it is not vulnerable". He included both the susceptibility of groundwater to possible contamination and the kinds and quantities of potential contaminants as essential parts of vulnerability assessment.

The U.S. Environmental Protection Agency (EPA) report on methods for assessing the sensitivity of aquifers to pesticide contamination recognized two types of the susceptibility of an aquifer to contamination resulting from the use of pesticides (US EPA, 1991). The term "aquifer sensitivity" was used for the intrinsic susceptibility of an aquifer to contamination based solely on the hydrogeological characteristics of an aquifer. "Aquifer vulnerability", a more comprehensive term, referred to the susceptibility of an aquifer to contamination that incorporates hydrogeological characteristics (aquifer sensitivity), land-use practices, and contaminant characteristics and loading (US EPA, 1991). The U.S. General Accounting Office (1991) used the term "hydrogeological vulnerability" for the intrinsic susceptibility of an aquifer to contamination and the term "total vulnerability" for vulnerability that is a function of hydrogeological factors, as well as of the land-use practices and contaminant loading.

Adams and Foster (1992) defined the vulnerability of an aquifer to contamination as being a function of (a) the inaccessibility of the saturated zone, in a hydraulic sense, to the penetration of contaminants and (b) the attenuation capacity of the strata overlying the saturated zone as a result of physicochemical retention or reaction of contaminants. These two factors interact with the mode of contaminant disposition, the magnitude of associated hydraulic loading, and the class of contaminants in terms of their mobility and persistence.

The Committee on Techniques for Assessing Groundwater Vulnerability of the (U.S.) National Research Council (1993) defined groundwater vulnerability to contamination as

"The tendency or likelihood for contaminants to reach a specified position in the ground water system after introduction at some location above the uppermost aquifer". But later in the text, the Committee also differentiated two general types of vulnerability: specific vulnerability (referenced to a specific contaminant, contaminant class, or human activity) and intrinsic vulnerability, which does not consider the attributes and behavior of specific contaminants.

## DEFINITION OF GROUNDWATER VULNERABILITY

The formulation and definition of what we understand by "groundwater vulnerability" and clarification of the concept of a vulnerability map is essential for the design, methods of cartographical representation, and compilation of vulnerability maps. Therefore, before we attempt to develop a generally acceptable vulnerability mapping procedure, the meaning of the term groundwater vulnerability must be carefully analyzed and defined.

For this book, we propose the following definition: "Vulnerability is an intrinsic property of a groundwater system that depends on the sensitivity of that system to human and/or natural impacts." We recognize that there may be more than one type of groundwater vulnerability. Therefore, we use in this book the term "intrinsic (or natural) vulnerability" for vulnerability defined solely as a function of hydrogeological factors--the characteristics of an aquifer and the overlying soil and geological materials. We further recognize that, in addition to intrinsic properties of a groundwater system, some users of vulnerability maps may also wish to include potential impacts of specific land uses and contaminants, which may prove detrimental--in space and time--to the present or future uses of the groundwater resource. For this concept, we use the term "specific (or integrated) vulnerability."

Development of a generally recognized and accepted definition of vulnerability does not imply a standardized approach to vulnerability mapping. Hydrogeological environments are much too diverse to be subjected to a standardized assessment. However, it is important to agree on a common base, i.e. definition of vulnerability, before we determine possible approaches to the assessment of these diversified conditions. We believe that the definition above will help remove the ambiguity and prolificity of the currently used terms and will help find a meaningful approach to vulnerability assessment and mapping.

Chapter 3.

# CHEMICAL AND BIOLOGICAL CONTAMINANTS AND THEIR SUBSURFACE BEHAVIOR

As mentioned in Chapter 2, earth materials have a certain capacity to remove some contaminants or reduce their concentration. This attenuation capacity, or "purification capacity", is a very important part of vulnerability assessment. It is not possible to adequately describe hydrogeological sensitivity to contaminants without evaluating the potential for contaminant attenuation. There are several broad groups of potential contaminants, each affected by different attenuation processes. Within each group there may be further classifications, unless a worst case situation is assumed.

Therefore, this chapter provides some insight into properties of the major contaminants and their impact on groundwater quality, and into the principal natural processes influencing their subsurface migration and attenuation. The chapter provides individuals intending to prepare or use groundwater vulnerability maps with a starting background concerning common chemical and biological contaminants and their behavior in the subsurface. It should be kept in mind that not all potential contaminants have been mentioned or discussed.

## GROUNDWATER CONTAMINATION

All groundwater contains dissolved solids; possesses physical characteristics such as odor, taste, and temperature; and sometimes contains naturally-occurring biological organisms such as bacteria. The natural quality of groundwater depends upon the physical environment and the origin and movement of water. As the water moves through the hydrological cycle, various chemical, physical, and biological processes change its original quality through reactions with soil, rocks, and organic matter.

Changes in groundwater quality are caused, directly or indirectly, by natural processes and human activities. Groundwater is degraded when its quality parameters are changed beyond their natural variations by the introduction or removal of certain substances. The degradation may impair the usefulness of water, but is not necessarily harmful to health.

The type, extent, and duration of induced changes of groundwater quality are controlled by the type of human influence; the geochemical, physical, and biological processes occurring in the ground (Table 1); and the existing hydrogeological conditions. The prediction of the effects of human interference requires knowledge of the position of the water table, the hydraulic gradient, the distance of wells or springs from hazardous activities, and the properties of the rocks, such as adsorption capacity and hydraulic conductivity. The subsurface movement of any contaminant is influenced by the moisture content and water balance in the unsaturated zone, the hydraulic gradient, and the water balance in the saturated zone. These parameters are controlled by the volume and flux of water in the system, which depend on climate, topography, and hydraulic conductivity.

**Table 1.** Natural processes controlling human influence on groundwater quality (modified from Langmuir, 1972) .

| Geochemical Processes | Physical Processes |
|---|---|
| Acid-base reactions | Advection/Convection |
| Adsorption-desorption | Dispersion |
| Complexation | Evaporation |
| Oxidation-reduction | Filtration |
| Solution-precipitation | Gas transport |
| | Radioactive decay |
| **Biochemical Processes** | **Biophysical Processes** |
| Cell synthesis | Filtration of pathogens |
| Organic decomposition | Transport of pathogens |
| Transpiration | |

Although natural processes may reduce the seriousness of groundwater contamination, many contaminants remain essentially unchanged after entering the groundwater body. Thus, their detrimental effect at a location may persist for years, decades, or centuries, because the average residence time of groundwater is measured in years; the comparable residence time of a contaminant in a surface water stream is days. Long periods are often required for contaminants to be removed from contaminated aquifers. And some aquifers or parts of aquifers may be damaged beyond repair (Zaporozec, 1981).

Since the International Symposium on Groundwater Pollution held in Moscow (USSR) in 1971 (Schoeller, 1975) worldwide scientific interest has focused more and more on the problems of groundwater quality and contamination. During recent years, the literature describing groundwater contamination and the fate and transport of contaminants in the subsurface has become overwhelming. Useful references providing a further detailed overview to the below discussion can be found in Canter et al, 1987; Johnson et al, 1989; and Matthess et al, 1985.

## PRINCIPAL CONTAMINANTS

Certain contaminants appear to predominate in groundwater. These contaminants are heavy metals, organic chemicals, and other substances such as fertilizer and pesticide constituents, bacteria, and viruses.

**Heavy Metals**

Heavy metals are usually defined as metals with densities greater than 5 $g/cm^3$. The sources of some metals are associated with natural processes, although many heavy metals result from human activities. The two most important natural processes that contribute heavy metals to natural water include chemical weathering and soil leaching. Heavy metals are used in various ways as raw materials for numerous industrial products or as catalysts in chemical processes. Some are constituents of fertilizers or pesticides, which are distributed over large areas by industrial or agricultural activities. Heavy metals may act as contaminants in soil, liquid, or gaseous wastes. Appreciable amounts of some heavy metals are set free by combustion of fossil fuels.

Wastewater, especially of industrial origin, often contains heavy metals. Solid waste dumps and residues from mining, ore processing, and smelting operations are common sources of higher local concentration of heavy metals in groundwater (Matthess, 1974).

Radioactive heavy metals may occur as fission products in connection with the processing and smelting of uranium ores; the production and reprocessing of nuclear fuel and explosives; the disposal of nuclear wastes; the escape of volatile radionuclides at nuclear power plants; and the various radionuclides used for medical or technical purposes.

Gaseous wastes, which are apt to propagate contamination within very short time periods and over wide areas, usually contain small quantities of some heavy metals. Examples would be lead from automotive traffic exhaust (Golwer and Schneider, 1973) and the fallout of the radionuclide $ruthenium^{106}$ (Aurand et al, 1971).

**Organic Chemicals**

Organic chemicals introduced into the earth's geochemical cycle contaminate groundwater. They are derived from various sources. A partial listing of approximately 200 classes of organic compounds observed in United States groundwaters (Table 2) indicate the wide range of composition and structure of the substances considered. Halogenated hydrocarbons (halogenated alkanes, olefins, benzenes, etc.) are among the most ubiquitous contaminants in groundwater.

Industrial and domestic flue gases and continental dusts contain organic chemicals such as hydrocarbons and halogenated hydrocarbons that may reach groundwater, dissolved in rain and seepage (percolating) water (Neumayr, 1981; Vrba, 1981; Zoeteman et al, 1981). Organic chemicals are present in municipal sewage and industrial liquid wastes. They may contaminate groundwater either by occasional leakage from sewers and canals or by intentional disposal into surface waters; by infiltration from septic tanks; by spreading of fertilizers and pesticides; or by injection into deep groundwater. Organic chemicals in solid wastes and sludges, when disposed of by controlled tipping (landfilling) or composting, are leached by rain and seepage water. Improper waste management frequently causes groundwater contamination (Jackson, 1980; van Duijvenbooden et al, 1981).

**Table 2.** Partial listing of classes of organic compounds detected in United States groundwaters (from Dunlap et al, 1983).

| Classes | Number per Class |
|---|---|
| Aliphatic hydrocarbons | 24 |
| Aromatic hydrocarbons | 27 |
| Compounds containing nitrogen and/or sulfur | 27 |
| Compounds containing phosphorous | 3 |
| Halogenated aliphatic hydrocarbons | 28 |
| Halogenated aromatic hydrocarbons | 6 |
| Halogenated phenols | 5 |
| Other halogen-containing compounds | 17 |
| Other oxygen-containing compounds | 53 |
| Phenols | 10 |

Organic chemicals that come into contact with rain water, surface water, seepage water, or directly with groundwater, are dissolved according to their solubility in water. Contaminated surface water may reach groundwater through artificial recharge of surface water or by bank-filtered river water. Additionally, contamination can occur through unused or abandoned wells, improper well construction, boreholes and excavations, and other hydraulic shortcuts (Zaporozec, 1981).

Pesticides used in agriculture and forestry for the control of detrimental organisms are mainly synthetic organic compounds. They are found in many groundwaters in very low trace concentrations (Jackson, 1980). Because of their widespread use and their persistence, groundwater contamination by pesticides is most probably by herbicides, insecticides, and nematicides. Pesticides have been found in many groundwater supplies. For example, during a groundwater monitoring program in the USA, at least 46 different pesticides were found to contribute to contamination of groundwater in at least 26 states (US EPA, 1990).

**Immiscible Organic Fluids**

Immiscible organic fluids, spilled on the ground by tanker accidents or released into the soil by leakage from pipelines or storage tanks, for example, may form separate organic phases that move and behave according to their individual physical properties. They may come into direct contact with groundwater or be dissolved in infiltrating water (Schwille, 1981). Substances such as gasoline, benzene, and most of the volatile chlorinated hydrocarbons will be more mobile than water in a given aquifer, whereas diesel fuel and heating oil will flow slower than water. Organic fluids that are lighter than water (gasoline, benzene, heating oil, etc.) may form bodies on the water table. The chlorinated hydrocarbons that are heavier than water (Table 3) may sink to the bottom of an aquifer. Since the 1970s,

**Table 3.** Densities of common hydrocarbon and halogenated hydrocarbon groundwater contaminants (at 20° C).

| | g/cm³ |
|---|---|
| Water | 0.9988 |
| Hydrocarbons | |
| Benzene | 0.879 |
| Gasoline | 0.725-0.785 |
| Diesel fuel | 0.82-0.86 |
| Halogenated Hydrocarbons | |
| Dichloromethane | 1.327 |
| Trichloromethane | 1.462 |
| 1,1,1-Trichloroethane | 1.337 |
| Dichlorobenzene | 1.306 |
| Tetrachloroethane | 1.598 |

volatile chlorinated hydrocarbons, as well as petroleum products (e.g., gasoline, kerosene, diesel, heating and motor oil) have been recognized as significant groundwater contaminants to an increasing extent.

**Other Contaminants**

Industrial and domestic flue gases contain carbon dioxide, sulfur dioxide, and, to a lesser extent, chlorine and fluorine, which can be detected in atmospheric aerosols. Solubilization in rain water allows introduction of such substances to groundwater.

The use of inorganic fertilizers directly increases the quantity of soluble salts in the soil. Fertilizers contain chlorides, nitrates, and phosphates of calcium, magnesium, ammonia, and sodium in varying proportions. A similar impact results from organic matter and soluble salts (especially chlorides and sulfates) in manures such as dung and liquid manure (George and Hastings, 1951; Schwille, 1962).

Domestic and industrial liquid wastes discharged to surface waters, infiltrated from septic tanks, spread as fertilizers, applied at land application sites (spray irrigation), or injected into deep hydrogeological structures are examples of groundwater contamination (Miller,

1975; Miller et al, 1977; van Duijvenbooden et al, 1981). Furthermore, uncontrolled leakage from sewers and canals must be taken into account (Vrba, 1981). There are numerous, well-investigated cases of groundwater contamination by disposal of municipal or industrial liquid wastes including arsenic, cyanides, nitrates, and phosphates (Jackson, 1980; Matthess, 1982; van Duijvenbooden et al, 1981).

The salinity of groundwater is often increased by the application of salt for snow and ice control on highways (Golwer and Schneider, 1973). This man-made deterioration of groundwater quality is important in smaller recharge areas crossed by numerous roads with intense salt application. Agricultural irrigation may increase salt and nitrate contents of groundwater due to evapotranspiration of the irrigation water, particularly when recirculated.

The injection of warm surface water or of groundwater that has been used for cooling purposes may change groundwater quality because of the increased capacity of warm water to dissolve rock constituents. The use of ground-water-based heat pumps for residential heating and cooling has resulted in thermal alteration of groundwater.

### Bacteria and Viruses

The contamination of groundwater by pathogenic bacteria and viruses has caused large outbreaks of waterborne diseases. The evaluation of case histories shows that outbreaks tended to happen in situations where downward-moving contaminated water by-passed the unsaturated zone by means of hydraulic shortcuts (Althaus et al, 1982). For example, the contaminant pathway is often poor well construction or design, or deteriorating well casings or grout (cement seals). The main contamination sources are nearby septic tanks, leaky sewer lines, sanitary landfills, waste oxidation ponds, and land application of wastewater.

The most important pathogenic bacteria and viruses that may possibly be transmitted in groundwater are: *Salmonella sp., Shigella sp., Yersinia enterocolitica, Yersinia pseudotuberculosis, Leptospira sp., Francisella tularensis, Dyspepsia coli*, enterotoxigenic *E. coli* (ETEC), Pseudomonades, *Vibrio sp., Legionella sp.*, infectious hepatitis virus, polio virus, coxsackie viruses, adenovirus, rotavirus, and Norwalk-like virus (Althaus et al, 1982; Gerba and Keswick, 1981).

## PROCESSES AFFECTING CONTAMINANT FATE AND TRANSPORT

The potential for groundwater contamination depends upon the attenuation of contaminants that takes place between the contamination source and the aquifer. The attenuation of contaminants as they travel through the soil zone, unsaturated zone, and groundwater system is affected by a variety of naturally occurring chemical reactions and biological and physical processes that often cause the contaminant to change its physical state or chemical form.

The principal reactions and processes are listed in Table 1. Their occurrence and intensity vary in the subsurface zones (Figure 2). The soil zone has the greatest variety and

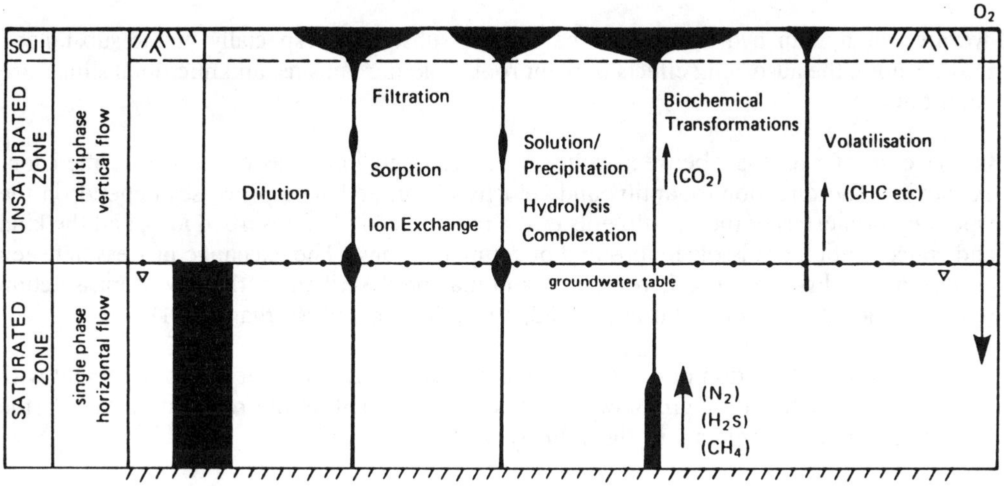


**Figure 2.** Processes causing contaminant attenuation in groundwater systems (from Foster, 1987; modified from Golwer, 1983). The thickness of the corresponding line indicates typically the relative importance of the process in the soil and above, at, and below the water table.

magnitude of natural processes, especially in the root zone, where significant amounts of chemicals are broken down by microorganisms or chemical and physical processes and taken up by plants. Less biological activity occurs in the unsaturated zone than in the soil zone, and the physical and chemical processes dominate here. The unsaturated zone's main feature is that it delays the arrival of contaminants to the water table. Fewer processes take place in the saturated zone where solution, dilution, and hydrodynamic dispersion are most effective in the attenuation of contaminants.

**Geochemical Processes Affecting Contaminant Transport**

The principal geochemical processes that alter the concentration of contaminants in groundwater are adsorption-desorption and solution-precipitation reactions, oxidation-reduction phenomena, and complexation.

***Adsorption-Desorption***
Many solid substances in the ground coming into contact with groundwater tend to release certain constituents into solution and to remove dissociated and non-dissociated components from solution by binding them to the surface of solid particles by intermolecular interchanges. This adsorption-desorption process is characterized by equilibrium between the quantity of a substance bound to an adsorbent and the quantity of this substance in solution. The relation shows that an increase in the concentration of a solution will raise the adsorbed quantity, and a decrease in concentration will result in desorption. Strong adsorbents in rocks include clay minerals, zeolites, iron and manganese hydroxides and

hydrates, aluminum hydroxide, and the organic substances, especially humic substances. Furthermore, the adsorbing effects of plant roots, microorganisms, and microbial slimes are important.

In the case of exchange between solute and adsorbed ions, the process is termed ion-exchange. The direction, quantity, and velocity of ion-exchange processes depend on the types and properties of the constituents of the rocks, the kind of adsorbed ions, and the kind and concentration of dissolved ions and of competing ions. The exchange process between adsorbed and dissolved ions is reversible and may be described by the law of mass action (Garrels and Christ, 1965; Matthess, 1982; 1990; Stumm and Morgan, 1981).

The continuous adsorption-desorption reactions cause retardation of the contaminant with respect to the surrounding groundwater, which is described by the retardation factor (the ratio of groundwater velocity to the velocity of the contaminant).

***Solution-Precipitation***

The concentration of dissolved contaminants in groundwater is the function of dissolution, degradation, and hydrolytic processes. Compounds are divided with respect to their dissolution behavior in water in electrolytes (salts, acids, bases) and non-electrolytes (polar and non-polar compounds). The ability of water to dissolve substances is increased by inorganic and organic acids and by an increase in temperature. Solution and precipitation are frequently controlled by pH and Eh. Electrolytic compounds mostly dissociate into ions. Polar organic compounds, such as sugar and alcohol, and gases form true solutions in which they occur as molecules. Nonpolar organic compounds such as mineral oil products and halogenated hydrocarbons are usually poorly soluble.

Along groundwater flow paths, dissolved materials may be precipitated when evaporation and transpiration increase their concentrations above the respective saturation limits (e.g., in arid climates) or when groundwaters of different chemical compositions are mixed. The addition of ions of the same species, especially of poorly soluble constituents, leads to precipitation when their solubility products are exceeded.

Precipitates usually remove other ions from solution during the process of precipitation of the ions of major concentration. This effect of co-precipitation is important for fixation of many heavy metals and radioactive substances in the ground, especially in iron and manganese hydrates. The contaminants are incorporated by isomorphous substitution of ions of similar size into the structure of the mineral forming (co-precipitation) or into one that has formed (replacement). The significant feature of co-precipitation and replacement processes is that the new solid phase is more insoluble than the original solid phase.

***Oxidation-Reduction***

Substances with solubilities dependent on pH and Eh may be precipitated by contact with groundwaters of different pH and Eh values (e.g., groundwater free of oxygen containing ferrous iron mixing with oxygen-bearing groundwater). Additionally, Eh conditions may change along groundwater flow paths. Oxygen-consuming processes, such as microbial degradation of organic matter, may give rise to oxygen-free reduction zones characterized by the presence of ferrous ion, manganese, ammonia, nitrite,.and sulfide; by the deficiency of nitrate; and by a diminished content or absence of sulfate (Schwille, 1976). In such reduction zones, heavy metals are precipitated as sulfides when sulfide ions are present. When groundwater flows into regions where oxygen supply exceeds the oxygen consumption, the reduced inorganic materials are oxidized and the poorly soluble hydroxides and oxides of iron and manganese are precipitated.

**Biochemical Processes Affecting Contaminant Transport**

Primary organic compounds are decomposed by microorganisms, which obtain from decomposition processes the carbon and hydrogen for their cell synthesis. The energy necessary for their metabolism is supplied by the degradation of substances rich in energy into simpler compounds, and finally into carbon dioxide and water. These reactions take place in both aerobic and anaerobic environments, although at a slower rate in an anaerobic environment. Under anaerobic conditions, the microorganisms receive necessary oxygen by reducing oxygen-bearing compounds, particularly nitrates and sulfates.

The directions of microbial reactions are controlled by the thermodynamic relations of the respective systems, but proceed under favorable ecological conditions much faster than as pure physical-chemical reactions. Microbial reactions are produced by autochthonic microorganisms that are adapted to local subsurface environments. An increase of nutrients by groundwater contamination produces an increase in microbial population density. Microbial reactions are hydrogeochemically important in the oxidation and reduction processes of the sulfur, nitrogen, iron, manganese, and carbon cycles (Matthess, 1982). Microbial degradation is disturbed by the presence of organic and inorganic substances that can inhibit metabolism or even kill the microorganisms. However, bacteria may become adapted to these substances (Knackmuss and Reineke, 1979).

**Physical Processes Affecting Contaminant Transport**

***Advection***

Advection is movement of contaminants caused by the flow of groundwater. Solutes or contaminants that do not react among themselves or with the solids of the aquifer are carried at the average rate of flow of groundwater. The rate of movement of a solute front, neglecting the tortuous flow paths in the porous medium and considering advective flow only, would be uniform along the entire front (Figure 3a). Therefore, solutes or contaminants appear to move as a straight line at the rate of groundwater flow.

Groundwater velocities in porous aquifers typically range from less than one mm/day to several m/day; however, usual range is from less than one m/day to a few m/day. Velocities above 10 m/day are restricted to very coarse sediments and high hydraulic gradients. In hard-rock aquifers, the groundwater flow velocities range from 0.3 m/day to 8000 m/day; in karstic aquifers up to 26 000 m/day. The spread of contaminants in highly fractured and karstic aquifers is much faster than in porous, non-indurated aquifers (Matthess and Pekdeger, 1981). The greater width of the interstices of fractured-rock aquifers enables the subsurface transport of suspended matter (particularly microorganisms, viruses, and substances giving rise to turbidity).

***Dispersion***

In natural porous materials that contain numerous, interconnected pores of different sizes, shapes, and orientations, solutes have a tendency to deviate from the anticipated flow paths. Because of varying velocities in this intricate network, a miscible fluid will spread gradually to occupy an ever increasing portion of the flow field into which it has been introduced.

This phenomenon is known as hydrodynamic dispersion ("mixing"), which can occur both in the direction of flow (longitudinal dispersion) and transverse to it (lateral dispersion). The front boundary of the body of contaminated groundwater appears "smeared" (Figure 3a). A portion of the solutes actually move ahead of what would have been predicted if only advection were considered.

Contaminated groundwater can be diluted by mixing with pure groundwater due to hydrodynamic dispersion until concentrations of the contaminants reach normal levels. The process of mixing causes the concentration of contaminants to decrease in time and with transport distance, whereas the volume of the contaminated plume increases.

***Retardation***

Another physical process governing migration of contaminants is retardation. In contaminant transport, there are a number of physical and chemical mechanisms that retard, or slow down, solute movement so that it may not move as fast as the advection rate would indicate. Solutes can be grouped in two broad classes: conservative and reactive (Fetter, 1988). Conservative solutes do not react with the soil and/or native groundwater (e.g., chloride). Reactive solutes can undergo chemical or biological change that reduces the solute concentration and will travel at a slower rate than conservative solutes. Figure 3b shows the impact of retardation on the movement of retarded solutes compared with nonretarded species. Retardation negates the effect of dispersion, and the solute front is again closer to the straight line. Retardation in a narrow sense is the effect of successive adsorption-desorption, which retards the contaminant transport relative to the water flow. In a broader sense, from a macroscopic view point, retardation may include dilution, filtration, chemical reaction, and biochemical transformation.

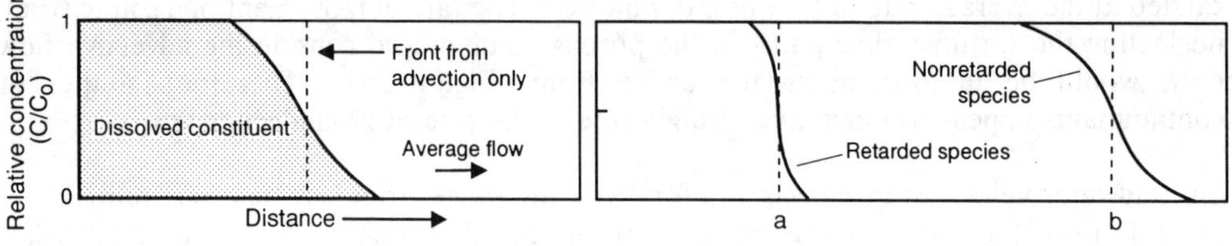


**3a.** Movement of a concentration front by advection and dispersion (after U.S. EPA, 1987).

**3b.**• Influence of retardation on movement of a solute front (after Fetter, 1988).

**Figure 3.** Migration of contaminants or solutes in groundwater.

***Filtration***
The filtration effect by soils and rocks is a complex physical and chemical phenomenon. It includes the removal of the larger particles by mechanical straining and the adsorption of smaller suspended particles (bacteria, flakes of iron hydroxide, etc.). The transport of suspended particles may be limited mechanically by the pore size and the size of the microorganisms or particles. Therefore, the mechanical filtering process in gravelly aquifers cannot be very effective due to the small diameters of flocculated iron hydroxides (10 $\mu$m), bacteria (0.2 to 5 $\mu$m), and viruses (0.25 to 0.02 $\mu$m) (Matthess and Pekdeger, 1981). More important is the particle accumulation on solid substance surfaces, which is affected by sedimentation, flow processes, diffusion, and interception (Matthess et al, 1991).

Sedimentation is very important for the accumulation of inorganic mineral in suspension (density about 2.5 g/cm$^3$), but not for microorganisms (particle size less than 5 $\mu$m, density about 1 g/cm$^3$). For particles with diameters of less than 1 $\mu$m (e.g., viruses) diffusion is very important, its effectiveness increasing with decreasing particle size. The processes of filtration have their minimum effectiveness at a particle size of about 1 to 5 $\mu$m, which corresponds to the size of most bacteria.

***Gas Transport***
Gas movement between groundwater and the atmosphere crosses two interfaces that separate the unsaturated zone from the groundwater and from the atmosphere. Gas movement in the unsaturated and saturated zones is due to diffusion, combined in the unsaturated zone with temperature and barometric changes and in the saturated zone with flow dispersion. The magnitude and efficiency of the oxygen supply from the atmosphere controls whether there are anaerobic conditions in the groundwater. The reverse movement of gas removes gaseous decay products, such as nitrogen, carbon dioxide, and volatile contaminants, from the groundwater.

Highly volatile substances, such as gasoline, benzene, and volatile chlorinated hydrocarbons, may preferentially escape from contaminated soil into ground air and into the free atmosphere (Fragiadiakis et al, 1979). But the volatility of a substance does not appreciably interfere with its concentration in groundwater. Thus the elimination effect of high vapor pressure is overestimated in groundwater (Zoeteman et al, 1981). Notwithstanding the slow release of gas from groundwater into the unsaturated zone, this process may result in increased carbon dioxide and hydrocarbon contents in the ground air, which allows measuring the extent of the contaminated zone (Albertsen and Matthess, 1978).

**Biophysical Processes Affecting Contaminant Transport**

Pathogens are passively entrained into and within groundwater. Extended propagation is only likely in large fissures and solution channels. However, even in these aquifers the unfavorable ecological conditions and the effect of antagonistic autochthonic organisms in the groundwater should eliminate these germs if the groundwater residence time is long (Althaus et al, 1982).

Lewis et al (1982) reviewed all published case histories of groundwater contamination by pathogens. They concluded that the horizontal travel distance of bacteria and viruses in the saturated zone is governed principally by groundwater flow velocity. In reported contamination incidents, the horizontal distance between the borehole or spring and the proven source of contamination was equivalent to no more than the distance traveled by groundwater in 20 days, despite the fact that pathogens are capable of surviving in the subsurface for much longer time; for example, Matthess et al (1988) mentioned up to 400 days. This observed restriction of the travel distance is due to the above mentioned filtration processes, which are mainly controlled by the geometrical features of subsurface voids (width, interconnection, etc.) and by the flow velocity (Matthess et al, 1988).

## Chapter 4.
# CLASSIFICATION AND REVIEW OF GROUNDWATER VULNERABILITY MAPS

Groundwater vulnerability maps are classified as interpretive groundwater protection maps, derived from general hydrogeological maps (Zaporozec, 1989). They differ from hydrogeological maps in that they do not show the elements of a groundwater system but the specific characteristics of these elements as they relate to vulnerability of groundwater.

## GENERAL CLASSIFICATION

Maps in general can be classified in many ways and no one classification is satisfactory for all purposes. Groundwater vulnerability maps are generally used for groundwater protection planning, decision-making, or management and belong to the category of environmental maps. Various criteria can be applied to the classification of environmental maps. According to Vrána (1984a), groundwater vulnerability maps are classified as special-purpose and applied environmental maps (Figure 4). The difference between special and applied maps is sometimes difficult to determine. Above all, the applied maps differ from the special ones by the selection of information presented.

## OVERVIEW OF VULNERABILITY MAP CLASSIFICATIONS

Vrána's proposal from 1981 for the classification of groundwater protection maps, based on the scale, content, and purpose of maps (Vrána, 1984a), can also be applied to vulnerability maps. Vrána classified maps as extra large (1:5 000 and larger)--special-purpose maps showing protection zones, point contamination incidents, etc.; large (1:10 000 to 1: 50 000)--multipurpose maps compiled for the areas of the great hydrogeological and water management importance; intermediate (1:100 000 to 1:500 000)--synoptical maps for water management and planning purposes at the regional and national level; small (1: 1 000 000 to 1:10 000 000)--general synoptical maps needed at the national and international level; and extra small (1:10 000 000 and smaller) maps for studies at the continental and global scales.

In Struckmeier's (1989) classification system, groundwater vulnerability maps are listed among problem-oriented, specialized maps derived from the general hydrogeological maps (Figure 5). He classified groundwater vulnerability maps as low-reliability, static maps with low level of information, usually constructed at a small scale and used for management and protection purposes.

Šarin (1989) included aquifer vulnerability maps among special-purpose hydrogeological maps showing single or limited data and mainly compiled at a large scale, regarded as an indispensable basis for urban planning.

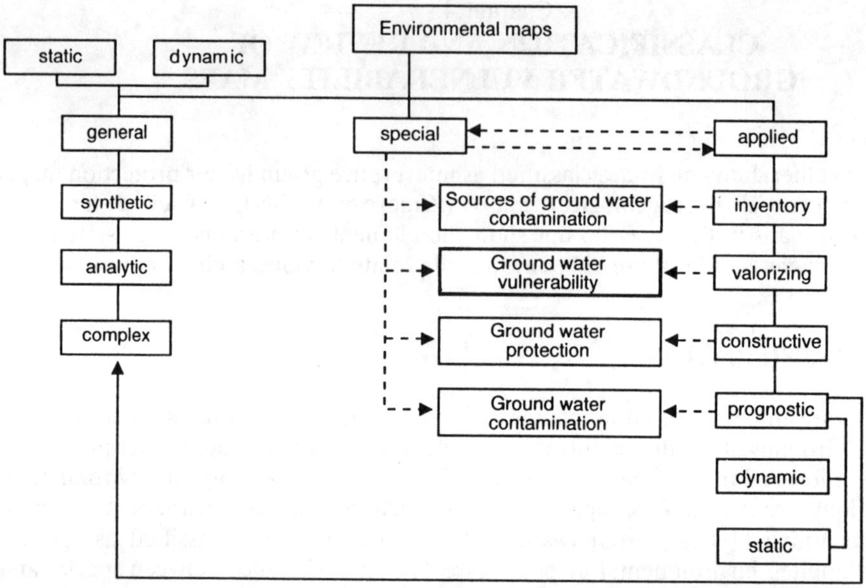


**Figure 4.** Classification of environmental maps (from Vrána, 1984a).

| level of information / possible use | **low** (scarce and heterogeneous data from various sources) | **advanced** (+ systematic investigation programmes, more reliable data) | **high** (+ hydrogeological systems analysis and groundwater models) |
|---|---|---|---|
| **reconnaissance** and **exploration** | **general hydrogeological map** (aquifer map) | **hydrogeological parameter maps** (map sets, atlasses) | **regional groundwater systems maps** (conceptual model representations) |
| **planning** and **development** | **map of groundwater resource potential** | **specialized hydrogeological maps** (planning maps) | **graphic representation derived from geographic information systems** (maps, sections, perspective diagrams, scenarios) |
| **management** and **protection** | **map of groundwater vulnerability** | | |
| possible use / parameters of representation | static → time-dependance → dynamic<br>low → reliability → high<br>low → cost per unit area → high<br>large ← area represented ← small<br>small → scale → large | | |

**Figure 5.** A classification system of hydrogeological maps (from Struckmeier, 1989).

Margat (1989) suggested that groundwater vulnerability maps should indicate the risk of contamination. The purpose of these maps is to provide information for decision-making on water and soil protection and initial diagnosis of impacts of accidental contamination. He sees the concept of maps of groundwater vulnerability to contamination as too complex and diversified one to be summarized in a single map and recommended to compile several maps showing more concrete and accurate information on vulnerability of groundwater to a specific contaminant (e.g., to nitrate resulting from agricultural practices).

In Freitag's classification (1989), vulnerability maps are listed in the category of serial maps that present various data sets in several maps either on one sheet or on several map sheets. According to Freitag, this type of maps are tools of geoscientific reasoning, spatial problem identification, and regional planning.

Zaporozec (1989) included groundwater vulnerability maps as a subdivision of groundwater protection maps, together with land suitability maps and special groundwater protection maps. According to their scale, he classified vulnerability maps into local (1:25 000 or less), regional (1:100 000 - 1:250 000), and national (1:1 000 000 - 1:2 500 000) maps. He recommended these maps as supporting tools for decision-making at all levels of governmental agencies responsible for the protection of public health and the environment.

Wang (1989) proposed classification of hydrogeological maps from the map user's viewpoint, and classified groundwater vulnerability maps as "evaluation" (or planning) maps. He considered vulnerability maps special-purpose maps, intended mainly for non-geoscientists.

A very sophisticated system of classifying, or better, typifying the use of hydrogeological maps is presented by Collin (1989). In his system, vulnerability maps are included among synthesis maps that offer synthetic information in a monothematic and multi-criterial way. Although the maps are monothematic, their preparation is difficult because numerous technical parameters and socio-economic constraints and criteria are to be considered. Collin sees vulnerability maps helpful for resource planning and decision-making.

In 1991, Vrba classified groundwater vulnerability maps according to scale, purpose, content, and graphical representation into four categories: specific, single-purpose (1:50 000 and less); specific, multipurpose (1:100 000 - 1:500 000); specific, general purpose (1:1 000 000 and more); and basic, showing the intrinsic vulnerability of groundwater (various scales).

## PROPOSED CLASSIFICATION SYSTEM

Based on the preceding overview and authors' experience, we recommend that groundwater vulnerability maps be placed within a classification scheme of environmental maps as one of many special interpretive maps. Scale, purpose, content, and graphical representation have decisive influence on vulnerability map classification. The recommended categories are shown in Table 4.

**Table 4.** Classification of ground water vulnerability maps.

| TYPE OF MAP | SCALE | PURPOSE AND CONTENT | GRAPHIC REPRESENTATION |
|---|---|---|---|
| GENERAL OVERVIEW SYNOPTICAL | 1:500 000 or more | General planning, decision-making, and setting policies in ground water protection on national or international level; educational purposes. Synthetical maps showing intrinsic vulnerability of ground water, local details are lost. | Mostly manually compiled, two-dimensional maps or atlases with explanatory notes; computerized maps are not frequent yet. |
| SCHEMATIC | 1:500 000 to 1:100 000 | Regional planning, ground water protection management and regulation, assessment of diffuse contamination problems. Most of local details are still lost, need to be followed by specific mapping. | Manually compiled maps or two- or three-dimensional computerized maps or atlases. |
| OPERATIONAL | 1:100 000 to 1:25 000 | District land-use planning and design of ground water protection programs. Analytical maps depicting vulnerability of ground water in areal extent in relation to the specific contaminant travel time. Field survey desirable. | Computerized digital two- or three-dimensional maps or manually compiled maps; cross sections and diagrams increase their usability. |
| SPECIFIC SPECIAL PURPOSE | 1:25 000 or less | Single-purpose and site-specific maps for local or city planning and well protection. Express local or site specific ground water vulnerability problems. Require set of representative data; site specific investigation is usually necessary. | Computerized digital two- or three-dimensional maps or diagrams (surface charts) and grid maps. |

Limitations on the use and application of vulnerability maps are given by the purposes for which they are compiled and by the scale that determines the map contents. The scale of vulnerability maps should be selected according to the purpose of the map, the character and complexity of hydrogeological conditions, and the accuracy required for problem solving. The scale plays an important role because it influences the accuracy and level of generalization of data as well as the values of parameters plotted.

For example, large-scale maps usually are special-purpose maps expressing contamination potential of a specific contaminant or a specific human activity. Such maps require representative, detailed data that are not always available, and therefore, a field investigation is necessary. On the other hand, the need for detail is much lower for the general overview (synoptical) maps showing intrinsic vulnerability at a national or international scale. Synoptical maps are mostly based on those characteristics of the elements of general hydrogeological maps that are related to groundwater vulnerability (e.g., lithology and permeability of rocks).

The map scale also determines the graphic representation of vulnerability. The large-scale specific maps, based on a large volume of data, increasingly are produced digitally or with the help of a geographical information system (GIS). Manual construction still is the preferred method for synoptical maps.

## REVIEW OF VULNERABILITY MAPS

The need to classify all sources of contamination and to define conditions of groundwater protection was advised by Vladimirskij as early as 1960 (Vladimirskij, 1960). He called attention to the fact that the maps showing sources of potential groundwater contamination are indispensable. He also suggested the classification of conditions and sources of contamination and preparation of a methodology for the construction of groundwater contamination potential maps, which were predecessors of groundwater vulnerability maps.

The first concepts and methodology of groundwater vulnerability maps were developed in the mid-1960s in Europe (Margat, 1968). In the late 1960s and the 1970s, the compilation of maps focused on groundwater contamination, protection, and vulnerability was well under way in several European countries (e.g., France, Czechoslovakia, and Germany). In the United States, Walker compiled a map of the contamination potential of aquifers in the state of Illinois (Walker, 1969), which is believed to be the first vulnerability map produced in the United States.

Some of the best known maps compiled in that period are summarized in a paper by Vrána (1984a) and include French (Albinet, 1970; Albinet and Margat, 1970; Tosan et al, 1975), Czechoslovakian (Olmer and Řezáč, 1974; Vrána, 1968), Polish (Kleczkowski et al, 1973; Macioszczyk and Plochniewski, 1979), Russian (Rogovskaya, 1976), Bulgarian (Antonov and Rajkova, 1978), German (Josopait and Schwerdtfeger, 1979), and Spanish (IGME, 1976) maps.

The map of aquifer vulnerability to contamination in France at the scale of 1:1 000 000, by Albinet (1970), was the first compilation to bear the title "vulnerability map". The map is based on the lithological composition of rocks. Six primary and nine secondary categories are defined according to the increasing contamination potential. For each category, permeability and groundwater velocity are determined. The categories are differentiated by colors. By hatching, supplemental information is shown: recharge areas, irrigated areas, and areas where aquifers are covered by semipermeable or impermeable layers at the surface. The direction of groundwater flow also can be deduced from the map. The French experience with groundwater protection maps promptly gained access to the South America as illustrated by Kreimer (1970) from Buenos Aires.

These examples demonstrate the first phase of the development of groundwater vulnerability maps. The authors of maps tried to solve the problems by constructing synoptical maps on a small scale, which covered the entire state territory. This method may be considered logical, taking into account the fact that the governmental authorities urgently needed such a basis to solve the most pressing and important tasks of groundwater protection at national or regional level (Vrána, 1984a).

The later phase of the development of map methodology was characterized by the transition to maps on medium and large scales. These maps have been developed because of the needs for groundwater protection of smaller territorial units. The most consistent approach to vulnerability mapping at these scales has been shown in France by the Bureau de Recherches Géologiques et Minières (BRGM). Lemaire and Martin (1973) prepared two maps of groundwater contamination potential at the scale of 1:100 000 (sheets Montpellier and Basse Valle de l'Aude). Lavie and Putallaz (1974) compiled four sheets of vulnerability maps at the scale of 1:50 000 in 1974. In 1976, with the sheet Lyon constructed at the scale of 1:50 000 by Beauduc et al (1976), a new edition of groundwater vulnerability maps has been initiated in which approximately four sheets appear every year. In the 1970s, a 1:200 000 series of vulnerability maps were started in the Czech Republic (Olmer and Řezáč, 1974).

Since the early 1980s a considerable number of vulnerability maps have been produced throughout the world, based mostly on aquifer vulnerability to contamination. Vulnerability mapping has been a major topic at two international meetings in the late 1980s. Several speakers at the 1987 International Conference on Vulnerability of Soil and Groundwater to Pollutants held in The Netherlands (van Duijvenboden and van Waegeningh, 1987) informed about the methodology and progress of vulnerability mapping in their countries (e.g., Breeuwsma and van Duijvenboden, Carter et al, Civita et al, Goosens and van Damme, Ostry et al, Sotorníková and Vrba, Subirana and Casas, and Wagner and Zomenis, all 1987). Also, at the 1989 International Symposium on Hydrogeological Maps as Tools for Economic and Social Development held in Germany (Struckmeier et al, 1989), the classification and methodology of vulnerability maps were discussed in many papers.

Vrána (1984b) reported on the progress of and methodology for the compilation of vulnerability maps in the Czech Republic, and demonstrated a new methodical approach on the example of the compilation of a synthetical map of groundwater vulnerability to acid rain.

In an attempt to achieve some national level of consistency in the United States, a standardized system for evaluating vulnerability of groundwater to contamination was developed (Aller et al, 1987). The objective of the system is to allow the evaluation of groundwater contamination potential of any area in the United States. The system, called DRASTIC (see list of acronyms), has two major parts: the designation of mappable units, termed hydrogeological settings, and the superimposition of relative numerical rating system.

Civita, Giuliano et al (1987) proposed the construction of vulnerability maps as a separate category of hydrogeological maps in the framework of a research programme through the Italian National Council for Research (CNR). Vulnerability maps are published at various scales (often 1:25 000 and 1:50 000) and are compiled with the objective of forecasting and preventing emergency situations in groundwater contamination.

In the Netherlands, vulnerability maps are produced for the entire country at the scale of 1:400 000 (Breeuwsma and van Duijvenboden, 1987) as special maps focused on portraying the characteristics of the soil and unsaturated and saturated zone relevant to the behavior of percolating contaminants. Maps are based on soil and geological maps, and are meant for survey at national level.

One county that has groundwater vulnerability maps at the general (1:1 000 000), schematic (1:200 000), and operational (1:40 000 and 1:100 000) scales, is Germany. As a part of the project "Grundwasservorkommen in der Bundesrepublik Deutschland" (Groundwater resources in the Federal Republic of Germany), carried out in 1977-78 by the Bundesanstalt für Geowissenschaften und Rohstoffe (Federal Institute for Geosciences), three maps of the republic were compiled at the scale 1:1 000 000 dealing with the most important characteristics of groundwater resources needed for the national and regional planning: availability, quality, and vulnerability to contamination (Aust et al, 1980; Vierhuff et al, 1981).

At the same time, systematical environmental mapping began at the scale 1:200 000, part of which was a hydrogeological map that included, among other data, vulnerability of groundwater to contamination (Josopait and Schwerdtfeger, 1979). Recently, the Federal Institute for Geosciences has begun a series of groundwater vulnerability maps at the scale 1:100 000 (J. Hahn, personal communication, 1994).

During 1980-85, hydrogeological mapping of the territory of the former German Democratic Republic was undertaken at the scale 1:50 000. The map series included five maps: hydrogeological map, aquifer map, pedological map, vulnerability map, and map of the Tertiary aquifer (H.-J. Voigt, personal communication, 1994).

In Sweden, various thematic maps on a scale 1:250 000 are derived from hydrogeological maps, including maps of groundwater vulnerability to surface contaminants (Engquist, 1989).

In the United Kingdom, several vulnerability maps have been produced. The English Midlands maps (1:100 000) indicate the vulnerability of selected aquifers to nitrate leaching (Lewis and Robins, 1989). Their purpose is to support interdisciplinary decisions and to provide a foundation for future policy on land-use modification and water protection. The National Rivers Authority of England and Wales published a national groundwater protection policy (1992), which will be supported by a series of regional intrinsic vulnerability maps at a scale of 1:100 000 (see example in Appendix B).

In the Czech Republic, a vulnerability map at the scale of 1:100 000 has been produced for the Czech Cretaceous Basin, the most important hydrogeological unit in the country (Sotorníková and Vrba, 1987). The map was constructed by applying vulnerability parameters (soils, unsaturated zone materials, aquifer transmissivity, and groundwater level fluctuations) to a hydrogeological base map.

During 1989-1991, a survey was conducted by the Wisconsin Geological and Natural History Survey to determine the status of vulnerability mapping practices in the United States (Zaporozec, 1993). Respondents viewed vulnerability maps as valuable derivative maps that show, quantitatively or qualitatively, certain characteristics of the subsurface environment that determine vulnerability of groundwater to contamination. They considered vulnerability maps to be useful for generalized application, and primarily useful as guidelines to the general public and governmental agencies at all levels as to the susceptibility of groundwater to contamination. Generally, vulnerability mapping per se has not begun in the United States until the mid-1980s. The survey of vulnerability mapping practices revealed that mapping methodologies vary greatly in objectives, scope, and presentation.

## GENERAL CONCEPT OF VULNERABILITY MAPS

The groundwater vulnerability map is a map showing a more or less subjective view of the capacity of the subsurface environment to protect groundwater, primarily in terms of water quality. It is subjective because the contents of the map must meet requirements or criteria of the map user. The essential purpose of the map is a subdivision of an area into several classes showing the differential potential for a specified purpose and use. Unlike geological maps, from which they are derived, groundwater vulnerability maps are time-dependent, constantly requiring updating to portray changes in both the characteristics of a groundwater system and the location and nature of potential contamination sources. Most vulnerability maps are constructed to evaluate the uppermost aquifer.

A vulnerability map should: provide the map user with the most accurate and informative assessment of aquifer sensitivity to human impacts; allow comparison of relative aquifer sensitivity to human impacts; allow comparison of relative aquifer sensitivity between different locations; and use all available, pertinent data in making the best possible interpretation.

Vulnerability mapping involves combining several thematic maps of selected physical resource factors into a groundwater vulnerability map that identifies different areas of the sensitivity of groundwater to natural and human impacts. Vulnerability mapping was defined by Bachmat and Collin (1987) as the technique of quantifying the assessment of vulnerability and displaying it (as a function of location and time) in a fashion that makes it useful and convenient for actual application in the decision-making process. In the view of Goossens and van Damme (1987), the vulnerability map is a map expressing the degree of risk for contamination of the groundwater in the upper aquifer by contaminants entering from the surface.

There are basically two approaches to vulnerability mapping, general and specific. The general or intrinsic vulnerability maps are used to evaluate the natural vulnerability of groundwater without context to a specific contaminant or a specific contamination source. Specific or integrated vulnerability maps (sometimes called land suitability maps) are used to evaluate the impact of a particular land use or a contamination source on groundwater; for example, vulnerability of a groundwater system to contamination by septic tanks. This approach implicitly includes characteristics of contaminants and the evaluation of the attenuation capacity for one or more contaminants.

Margat (1991) suggested that general vulnerability maps are useful for sensitizing planners to groundwater protection issues at the beginning of a regional planning process. At a later stage in the planning process, specific vulnerability maps taking into account the contamination risk are preferable.

A vulnerability map is based on the assessment and display of several parameters, which vary over regions as a function of the physical environment. A number of parameters have been proposed by authors of various maps. The individual parameters are discussed in Chapter 5. However, the principal parameters are associated with:

(a) The hydrogeologic framework--characteristics of the soil, unsaturated zone, and aquifer materials, and depth to groundwater.

(b) The groundwater flow system--the direction and velocity of the groundwater flow and topography.

(c) The climate--amount of recharge to groundwater.

Chapter 5.

# ASSESSMENT OF GROUNDWATER VULNERABILITY

The assessment of groundwater vulnerability and methods and techniques of its graphical and numerical representation are essential in the compilation of vulnerability maps. All groundwater, with possible exception of fossil water that has not been new part of the hydrological cycle and deep-seated brines, is vulnerable to various degrees. Vulnerability of groundwater is a relative, non-measurable, dimensionless property. The accuracy of its assessment depends above all on the amount and quality of representative and reliable data. Such data are not always available. The lesser the amount of data and the knowledge of the groundwater system, the lesser the reliability of the assessment of groundwater vulnerability.

The selection of assessment methods and data requirements depends on the purpose of the assessment and criteria given by the map user. Vulnerability is most often assessed in terms of water quality, and the assessment is made of the uppermost aquifer. Vulnerability assessment of deeper aquifers is less frequent.

## ASSESSMENT OF INTRINSIC VULNERABILITY

Vulnerability maps cannot be produced without consideration of the individual factors that determine the homogeneity of the areas under study and their capacity for attenuating contaminants. The intrinsic (natural) vulnerability map is based on the assessment of various natural factors or attributes, such as soils, the unsaturated zone, aquifer properties, and recharge rate, that enter into the determination of the vulnerability of groundwater. The most common attributes and their parameters are listed in Table 5.

When assessing groundwater vulnerability, the attributes or their parameters may be assigned different weights and rating according to their considered importance for the vulnerability assessment. Despite the differences in opinion as to the weight and rating of the individual attributes, it is generally recognized that groundwater vulnerability can be assessed only when the basic parameters of the attributes mentioned above are known.

### Attributes of Primary Importance

The principal attributes of intrinsic groundwater vulnerability include recharge, soil properties, and the characteristics of the unsaturated and saturated zone (see Table 5).

#### *Recharge*

Recharge, as used in this report, is the amount of water passing through the unsaturated zone and into an aquifer during a specified period of time. Recharge is usually expressed as annual net recharge. The amount and quality of recharge significantly affects the physical and chemical processes in the soil-rock-groundwater system.

**Table 5.** Intrinsic and specific attributes of groundwater vulnerability and their parameters (modified from Vrba, 1991).

| ATTRIBUTE / PARAMETERS | INTRINSIC (NATURAL) | | | |
|---|---|---|---|---|
| | Primary importance | | | |
| | Soil | Unsaturated (vadose) zone | Aquifer (saturated zone) | Recharge |
| PRINCIPAL | Texture<br>Structure<br>Thickness<br>Content of soil organic matter<br>Clay mineral content<br>Permeability | Thickness (related to ground water table)<br>Lithology (with regard to consolidation and stratification of rock)<br>Travel time of water | Lithology (with regard to consolidation and stratification of rock)<br>Thickness<br>Effective porosity<br>Hydraulic conductivity<br>Ground water flow direction<br>Age and residence time of ground water | Net annual recharge rate<br>Annual precipitation |
| SUPPLEMENTAL | Cation exchange capacity<br>Reduction and sorption capacity<br>Carbonate<br>+Bulk density<br>+Full soil water capacity<br>+Water uptake by plant roots<br>+Nitrogen transfer reactions | Weathering rates<br>Permeability | Hydraulic inaccessibility<br>Storage capacity<br>Transmissivity | Evaporation<br>Evapotranspiration<br>Air temperature |

+ Mostly assessed only when ground water vulnerability to diffuse contamination from agricultural activities is studied.

**Table 5.** (continued)

| ATTRIBUTE / PARAMETERS | INTRINSIC (NATURAL) | | | SPECIFIC |
|---|---|---|---|---|
| | Secondary importance | | | |
| | Topography | Underlying geological unit of aquifer | Contact with surface and sea water | |
| PRINCIPAL | Slope variability of land surface | Permeability<br>Structure and tectonics<br>Potential recharge/ discharge | Gaining/loosing stream<br>Evaluation of potential bank infiltration<br>Interface of salt/ fresh water in coastal areas | Land use<br>Natural: forest, meadow, non-farm land<br>Man-made: farm land, land under industrial activities, rural or urban settlements<br>Population density<br>Travel time of contaminants in the unsaturated zone<br>++Attenuation capacity of the soil, unsaturated zone, and aquifer in relation to specific contaminants |
| SUPPLEMENTAL | Vegetative cover | In case of confined aquifer the same parameters apply for the underlying unit of aquifer | | Residence time of contaminants in aquifer<br>Transport characteristics of contaminants - distribution coefficient (Kd) values, persistence (half-life)<br>Artificial recharge<br>Irrigation<br>Drainage |

++ For definition, see Glossary.

Recharge, as an attribute of the primary importance, should always be considered in the assessment of groundwater vulnerability, particularly on medium- and small-scale maps or when the vulnerability to drought is assessed. Recharge may be evaluated on the basis of field measurements, derived from the water balance equation, or estimated with the help of aerial photographs or satellite imagery. Also needed are the climatic data, such as precipitation, air temperature, and evaporation, that significantly influence the amount of recharge.

Recharge is frequently used in the U.S. vulnerability maps (Johnston, 1988) and its importance is regarded rather highly (e.g., weight 4 in DRASTIC system, Aller et al, 1987). Several European authors incorporated recharge into vulnerability assessment (Breeuwsma et al, 1986; Civita, 1990a; Josopait and Schwerdtfeger, 1979; Marcolongo and Pretto, 1987). Andersen and Gosk (1987) used recharge in their concept of the "restoration capability of an aquifer". The capability is defined as the volume of water contained in the aquifer ($m^3$) divided by the rate of recharge ($m^3$/year). Recharge was also applied as an attribute in mapping the groundwater vulnerability to contamination in the Munich-Harlaching area (Hafen et al, 1989). The potential annual recharge was used as one of the main attributes when the sensitivity of European aquifers to acid deposition was assessed (Holmberg et al, 1987).

***The Soil***

The upper unconsolidated layer of the Earth's crust is commonly regarded as one of the principal natural factors in the assessment of groundwater vulnerability. The main soil parameters related to vulnerability include texture, structure, thickness, and the content of organic matter and clay minerals (see Table 5). Other soil parameters, such as soil moisture, should be evaluated when available. If developed, the soil usually forms a continuous layer but the spatial variability of its physical, chemical, and biological properties is great. Therefore, any generalization of soil parameters should be done with great care. The soil has an important attenuation function (Zaporozec, 1985) and is a critical attribute when groundwater vulnerability to diffuse contamination sources (fertilizers, pesticides, acid deposition) is assessed.

The soil has a specific position among the groundwater vulnerability attributes because it itself is very vulnerable. The soil's function as a natural protective filter in the retardation and degradation of contaminants can be damaged relatively easy. The damage may lead to the loss of its control over groundwater quality. Therefore, the soil properties assessment should always take into consideration whether the soil in the area under study is in natural conditions or under stress from agricultural activities, acid deposition, etc.

Although soil parameters are usually available from various agencies, it is not complicated or costly to obtain them by field measurements or from published material. Aerial photographs and satellite imagery are helpful and can frequently be used for the evaluation of soil parameters if supplemented with field checks.

***The Unsaturated Zone***
This zone is very important in the protection of groundwater especially in hilly and mountainous regions and in areas where the soil profile is not well developed. Then the character of the unsaturated zone and its potential attenuation capacity decisively determine the degree of groundwater vulnerability. If this zone is composed of low permeable rocks, it creates a confining layer for the underlying aquifers and reduces significantly their vulnerability.

The main parameters included in the assessment are the thickness, lithology, and vertical permeability (see Table 5). The thickness of the unsaturated zone depends on the position of the water table, which is not stable and fluctuates frequently. For this reason, an analysis of groundwater level fluctuations should be included in vulnerability assessment. The minimum thickness of the unsaturated zone is given by the highest elevation of the fluctuating water table for the period of record. Supplementary parameters may include the degree of weathering of the upper part of the unsaturated zone.

It is more difficult and costly to obtain the necessary data on the unsaturated zone than on the soil. Drilling of the exploratory and monitoring boreholes, field and laboratory measurements and observations, and isotope studies are desirable for the assessment of the unsaturated zone. Aerial photographs and satellite imageries yield less valuable information, particularly when the unsaturated zone is stratified and thick.

***The Saturated Zone***
An aquifer (the saturated zone) is not a homogenous unit but a heterogenous system. Its vulnerability varies spatially and with depth. The aquifer vulnerability should be differentiated horizontally (recharge and discharge areas), vertically (oxidation, intermediate, and reduction zones), and according to the existing groundwater flow systems of varying geographical extent (local or regional) and depth (shallow or deep). The definition of semiconfined, confined, and unconfined conditions is quite important and must always be considered when assessing aquifer vulnerability.

The main parameters for assessment of aquifer vulnerability include the aquifer nature and geometry, porosity, hydraulic conductivity, storage properties, transmissivity, and groundwater flow direction (see Table 5). The importance of hydraulic conductivity is especially emphasized. Obtaining representative data on aquifer parameters is more expensive and technically demanding in comparison with data on the soil and unsaturated zone; exploratory and monitoring boreholes, hydraulic tests, field and laboratory analyses, and related data bases are needed to perform a reliable vulnerability assessment.

**Attributes of Secondary Importance**

Natural attributes of secondary importance include: topography, surface water, and the nature of the underlying unit of the aquifer (see Table 5). Their importance for vulnerability assessment varies with the area. Depending on the natural conditions, the importance may be greater in some areas such as: flat recharge areas, bank infiltration from a surface stream into a shallow aquifer, groundwater contact with the underlying strata of high ion-exchange capacity; and smaller in others (steep-slope recharge areas, low level ion-exchange or sorption capacity of the underlying strata). An important attribute is topography, which influences recharge, soil development, and groundwater flow and velocity.

A useful method of supporting assessment of groundwater vulnerability can be the application of environmental radioactive isotopes to determine the age and residence time of groundwater (Custodio, 1990). Particularly $^{18}O$, $^{2}H$, radioactive $^{3}H$, $^{13}C$, $^{34}S$, and $^{14}N$ are stable isotopes and are subjected only to small changes. However, Custodio pointed out certain limitations in the use of $^{34}S$, $^{18}O$, and $^{14}N$.

## ASSESSMENT OF SPECIFIC VULNERABILITY

Specific vulnerability of a groundwater system is mostly assessed in terms of the risk of the system becoming exposed to contaminant loading. Specific vulnerability of groundwater is assessed using methods of different levels of sophistication. Some authors call for a maximum simplification of specific vulnerability assessment (Andersen and Gosk, 1987).

### Factors Affecting Specific Vulnerability

In comparison with the assessment of natural vulnerability, which is based mostly on the static intrinsic parameters of the soil-rock-groundwater system, the dynamic and variable parameters are included in the assessment of specific vulnerability (see Table 5). The contaminant's travel time in the unsaturated zone and its residence time in an aquifer are often introduced.

The important parameter in the assessment of specific groundwater vulnerability is the attenuation capacity of the soil, of the unsaturated zone, and of the aquifer with respect to the properties of individual contaminants. The attenuation capacity of these media with respect to a particular contaminant can be exceeded or reduced over time, which results in a changed vulnerability of the groundwater system to that contaminant. A special approach is required for persistent and mobile contaminants. In their case the role of attenuation processes in the soil and the saturated zone is minimal and the aquifer's vulnerability depends on its thickness and permeability (residence time of the contaminant). The aquifer has to cope with the persistent contaminant on its own and its vulnerability depends mainly on the amount of water stored in the aquifer and the net recharge. Both of these parameters control the dilution of the persistent contaminant in groundwater, which is the only important attenuation process available in the aquifer system.

Major attributes involved in assessing specific groundwater vulnerability include: land use (human impact) and population density. There is a fundamental difference between areas with land under human stress (agriculture, industry, settlements, acid deposition) and areas where natural landscape with natural vegetation predominates (forests, uncultivated meadows, unpopulated mountainous regions). The more densely an area is populated, the greater the potential and real contaminant load on the groundwater system.

### Single-Purpose Assessment

The single-purpose assessment is the simplest concept of the specific groundwater vulnerability assessment. The vulnerability is evaluated with respect to only one type of contaminant or one group of contaminants having similar properties. The attenuation capacity of the soil, unsaturated zone, and aquifer for a particular type of contaminant and its properties, travel time, and residence time are assessed and mapped. Usually a single map suffices for portraying single-purpose specific groundwater vulnerability.

An example at the international level is the assessment of the sensitivity of European aquifers to acid deposition (Holmberg et al, 1987). Good examples at the national and regional levels are the maps of vulnerability to nitrate leaching for selected major aquifers in the United Kingdom (Carter et al, 1987) and for southern Ontario aquifers in Canada (Ostry et al, 1987). Groundwater vulnerability to nitrate--a highly mobile and stable contaminant, especially in the aerobic conditions of shallow aquifers--is based on the assessment of properties of the soil-rock-aquifer system. The travel time of nitrate is not considered in these maps.

Single-purpose specific vulnerability of groundwater is often assessed at the local scale with respect to point sources of contamination. Contaminants from point sources often enter the groundwater system under the soil profile (leaking underground tanks, septic tanks, etc.). In such cases the role of soil as an attenuation medium is nil, which considerably increases the aquifer's vulnerability.

### Multi-Purpose Assessment

This specific assessment of groundwater vulnerability includes an assessment of two or more contaminants or groups of contaminants and mapping at different scales. Examples of multipurpose specific assessment are: heavy metals in mining areas, nutrients and pesticides in agricultural areas, and pathogens and microorganisms in rural areas. As in the case of single-purpose vulnerability, the soil in many cases cannot be included in the assessment. Recharge, time of travel, and contaminant movement through the underground system are widely used for assessing specific vulnerability in order to illustrate the diversity of contaminants and variety of their individual properties.

It is difficult to assess and portray contaminants and their properties on a single map as groundwater vulnerability differs for different contaminants. Specific multipurpose vulnerability of groundwater, therefore, can be portrayed several map sheets using transparent overlays, superimposed maps, or atlases.

### Assessment of Specific Groundwater Vulnerability on a Synoptical Scale

Many authors express doubt whether specific groundwater vulnerability can be assessed and depicted on small-scale synoptical maps, pointing out that to assess and map vulnerability of groundwater to a broad range of contaminants of different properties is too complicated. The idea of a "general contaminant" or "universal contaminant" is not regarded as realistic.

Andersen and Gosk (1987) emphasized that "applicability of a general type, ready-made vulnerability map valid for all geological, hydrological, and hydrochemical situations, not considering pollutant type and pollutant scenario, is very limited."

Margat and Suais-Parascandola (1987) are optimistic about the assessment and mapping of multipurpose specific vulnerability. They point to the progress in computer graphics coupled with cartographic data bases, which opens up new prospects in the assessment of groundwater vulnerability to contamination, particularly in the synthetical visualization of the spatial variations of an object, both multiparametric and relative to multiple criteria.

## METHODS AND TECHNIQUES OF GROUNDWATER VULNERABILITY ASSESSMENT

The evaluation of groundwater vulnerability (as shown in the preceding section) should be made case by case, particularly if we want to take into account all of the following: the chemical and physical characteristics of every single contaminant (or of a group of contaminants); the type of contaminant source (point or diffuse); and the quantity, means, and rates of contaminant applications (Andersen and Gosk, 1987; Bachmat and Collin, 1987; Foster and Hirata, 1988). Such an approach is scientifically valuable and adequate for the assessment of specific vulnerability of groundwater to contamination from a point contaminant source in a small area (LeGrand, 1983; Seller and Canter, 1980). However, it is quite impractical for the assessment of intrinsic vulnerability of large areas, prepared for contamination prevention and aquifer protection planning. In the last twenty years, a number of techniques have been developed for these purposes.

The parameters and methods used for vulnerability assessment are listed in Tables 5 and 6. Parameters include among others soil characteristics, hydrological features of the saturated and unsaturated zone, net recharge, depth to water, and permeability of aquifers. Some authors also add other parameters, which are much more difficult to collect and often hardly available in some localities.

These techniques vary according to the following factors: the physiography of the area under study, the quantity and quality of data, and the purpose of the studies. In general, techniques can be subdivided into two distinctive classes: universal--that may be used for any physiographical scenario, or local--that may be used for only one particular area. However, with respect to type, the techniques can be grouped into three basic groups: hydrogeological setting methods, parametric methods, and analogical relation and numerical model methods. These methods are summarized in Table 6.

### Hydrogeological Complex and Setting Methods

The hydrogeological complex and setting (HCS) methods of assessing groundwater vulnerability involve the comparison of a subject area to criteria judged to represent conditions found to be vulnerable in other areas. Generally, a hierarchical system of two or more classes are established to span the continuum of vulnerability. These widely used methods evaluate vulnerability of hydrogeological complexes and settings, generally using

**Table 6.** Main methods for the assessment of intrinsic vulnerability of groundwater (from Civita, 1993).

| METHOD | | BASIC PARAMETERS | | | | | | | | | | | | | |
|---|---|---|---|---|---|---|---|---|---|---|---|---|---|---|---|
| REFERENCE | TYPE | | | | CHARACTERISTICS OF SOILS | | | | | | | | | | |
| | | PRECIPITATION RATE & CHEMICAL COMPOSITION | TOPOGRAPHIC SURFACE SLOPE VARIABILITY | STREAMFLOW NETWORK DENSITY | THICKNESS, TEXTURE & MINERALOGY | EFFECTIVE MOISTURE | PERMEABILITY | PHYSICAL & CHEMICAL PROPERTIES | AQUIFER CONNECTIONS TO SURFACE WATER | NET RECHARGE | CHARACTERISTICS OF THE UNSATURATED ZONE | DEPTH TO WATER | WATER LEVEL CHANGES | HYDROGEOLOGICAL FEATURES | AQUIFER HYDRAULIC CONDUCTIVITY |
| Albinet & Margat (1970) B.R.G.M. (1976...) | HCS | | | | | | | | ● | | ● | ● | | ● | ● |
| Vrana (1968) Olmer & Rezac (1974) | HCS | | | | | | | | | | ● | | | ● | |
| Fenge (1976) | RS | | | | ● | | | | | ● | ● | ● | ● | ● | ● |
| Josopait & Schwerdtfeger (1979) | HCS | | | | | | | | | ● | ● | ● | | ● | ● |
| Zampetti (1983) Fried (1987) | AR | | | | | | | | | | ● | ● | | | |
| Villumsen et al (1983) | RS | | | | ● | | | | | | ● | ● | ● | ● | ● |
| Haertle (1983) | MS | | | | | | | | | | ● | ● | | | |
| Vrana (1984b) | HCS | ● | | | ● | | | | | | ● | | | ● | |
| Subirana Asturias & Casas Ponsati (1984) | HCS | | | | | | | | ● | | ● | ● | | ● | ● |
| Engelen (1985) | MS | | | | | | | | ● | | ● | ● | | ● | |
| Zaporozec (1985) | RS | | | | ● | ● | ● | ● | | | ● | ● | | ● | |
| Breeuwsma et al (1986) | HCS | | | | ● | ● | ● | ● | ● | ● | ● | ● | | | ● |
| Sotornikova & Vrba (1987) | RS | | | | | | ● | | | | | ● | ● | ● | |
| Ostry et al (1987) | HCS | | | | ● | | | ● | | | | ● | | ● | |
| Ministry Flemish Comm. (1986) Goossens & Van Damme 1987) | MS | | | | ● | | | ● | | | | ● | | ● | |
| Carter et al (1987) Palmer (1988) | MS | | | | ● | | ● | ● | | | | | | ● | |
| Marcolongo & Pretto (1987) Method 1 | RS | | | | ● | | | | ● | ● | ● | | | | |
| Marcolongo & Pretto (1987) Method 2 | AR | | | | | ● | | | | ● | ● | ● | | | |
| GOD - Foster (1987) | RS | | | | | | | | | | ● | ● | | ● | |
| Schmidt (1987) | RS | | | | ● | | | | ● | | ● | ● | | | |
| Trojan & Perry (1988) | PCSM | ● | ● | | | | ● | | | ● | ● | ● | | ● | |
| Civita in Benacchio et al (1988) | HCS | | | | | | | | ● | | ● | ● | | ● | ● |
| DRASTIC - Aller et al (1987) | PCSM | | ● | | ● | | | | | ● | ● | ● | | ● | ● |
| SINTACS - Civita (1990a) | PCSM | | ● | ● | ● | | | | ● | ● | ● | ● | | ● | ● |

Explanation: AR - analogical relations, HCS - hydrogeological complex and setting, MS - matrix system, PCSM - point count system model, RS - rating system.

an overlay cartographic method (Albinet and Margat, 1970; Antonov and Rajkova, 1978; Aust et al, 1980; BRGM, 1973; 1975-1979; 1976...; Francani and Civita, 1988; IGME, 1979; Olmer and Řezáč, 1974; Rogovskaya, 1976; Subirana and Casas, 1984). These methods belong to the category of universal type systems, therefore, they are suitable for large areas with a variety of hydrogeological, hydrostructural, and morphological features. Hence, they are best suited to produce thematical maps at a medium to large scale, or to cover entire national territories. The vulnerability assessment is given only in qualitative terms (Civita, 1990b).

**Parametric System Methods**

The second group includes a variety of parametric systems that may be divided into:

(a) matrix systems (MS),
(b) rating systems (RS), and
(c) point count system models (PCSM).

The overall procedure for the various parametric systems is the same. The construction of a parameter system begins with the selection of factors (parameters) judged to be representative to assess the vulnerability of groundwater. Each has a defined natural range, which is subdivided into discrete hierarchical intervals (for example: 0-5m, 5-10m, and 10-20m to groundwater). Each interval is assigned a value reflecting the relative degree of sensitivity to contamination. Values usually are on a scale from 1 to 10, with 10 being the most sensitive.

***Matrix Systems***
The matrix systems are always suitable for local use. They are based on a limited number of carefully chosen parameters. A system selected for the Flemish region of Belgium (Goossens and van Dame, 1987; Ministry of the Flemish Community, 1986) includes three types of covering (soil), two intervals of depth to water, and four aquifer types. Also the system presently in use in some areas of central England under the jurisdiction of the Severn-Trent Water Authority (Carter et al, 1987; Palmer, 1988) is based on a matrix using four types of soil leaching characteristics (texture and physical and chemical properties) with three aquifer settings (Figure 6). Other interesting matrix systems have been used by Engelen (1985), Haertlé (1983), and Josopait and Schwerdtfeger (1979).

Adams and Foster (1992) recommended to retain hydrogeological variables in vulnerability assessment rather than to rank the geological parameters comprising vulnerability. They

| Aquifer Classification Type | Soil Leaching Class | | | |
|---|---|---|---|---|
| | 1 | 2 | 3 | 4 |
| 1 | EXTREME | HIGH | MODERATE | LOW |
| 2 | HIGH | MODERATE | LOW | LOW |
| 3 | LOW | LOW | LOW | LOW |

**Figure 6.** The matrix system used for groundwater vulnerability classification of Map 5 - Lichfield, England (from Palmer, 1988).

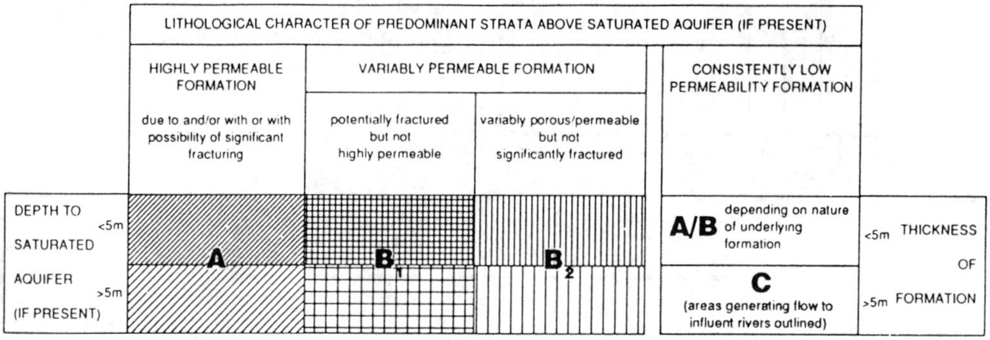


**Figure 7.** Example of zones based on aquifer vulnerability (from Adams and Foster 1992).

recognized three main classes of vulnerability (Figure 7), based on the permeability of the strata overlying an aquifer and the depth to water. Influent rivers are an additional consideration, where these have significant extension upstream of their influent sections on class C areas.

***Rating Systems***

The simple rating systems are largely derived from LeGrand's systems (1964 and 1983). A fixed range is given to any parameter that is judged necessary and adequate for vulnerability assessment. The range is properly divided, according to the variation interval of each parameter. The sum of rating points gives the required evaluation for any point or area. The final numerical score is divided into segments (from minimum to maximum) expressing a relative vulnerability degree. Many parameters are used in the rating systems. Some authors primarily use soil characteristics (see Table 6); for others, the hydrogeological and hydrological parameters are more important.

The rating systems are an extension of the hydrogeological setting methods in that they involve calculation of a rating or numerical score for each hydrogeological setting present in the area to be assessed. The rating schemes are based upon the assumption of a generic contaminant; they are not intended to be specific to any particular contaminant.

This type of system has been used by Fenge (1976) for the Saanich Peninsula in British Columbia, Canada; Marcolongo and Pretto (1987) for a representative area of Veneto Plain, Italy; Sotorníková and Vrba (1987) for a part of the Czech Basin; and Villumsen et al (1989) for Djursland Peninsula in Denmark. Zaporozec (1985) and Schmidt (1987) have also proposed similar techniques in Wisconsin, USA, vulnerability studies.

One of the more interesting rating systems, due to the simple and pragmatic structure, has been proposed by Foster (1987), with the acronym GOD (Figure 8). Equally interesting, especially for the large plains like the Po valley, is the system proposed by Marcolongo and

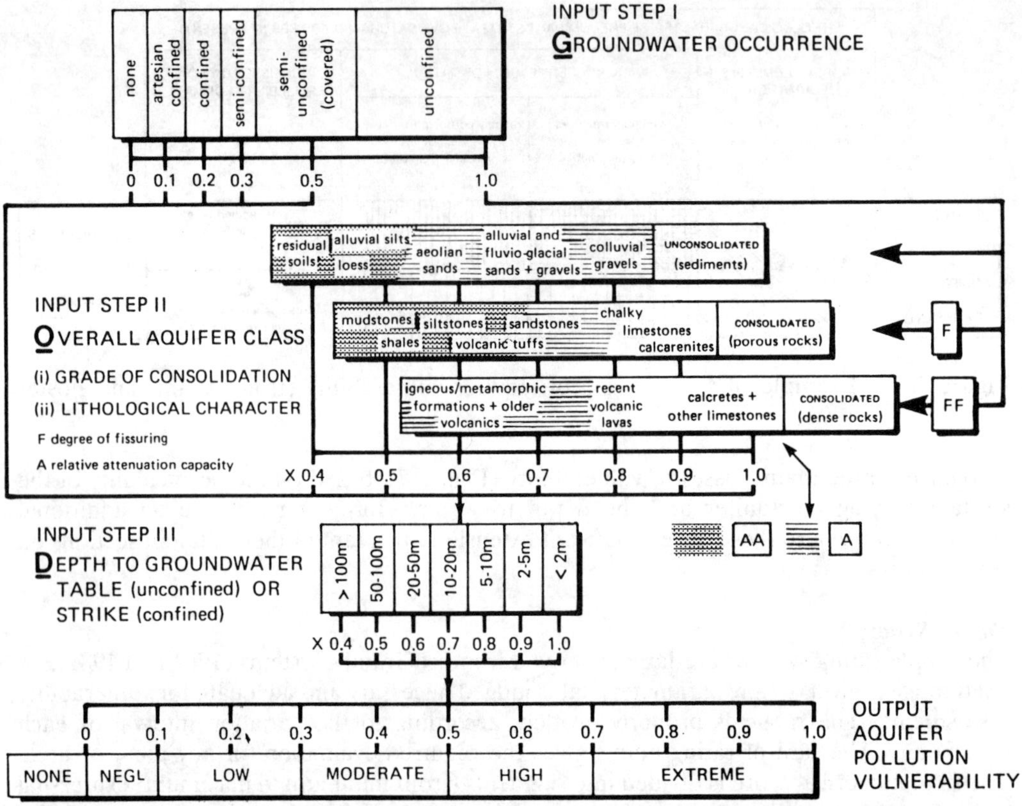


**Figure 8.** GOD empirical system for the rapid assessment of aquifer contamination vulnerability (from Foster, 1987). *Editorial Note: Corrections received from the author* - Step I: substitute "overflowing" for "artesian confined"; Step II: title should be "Overlying Lithology"; Output: omit "none".

Pretto (1987), which takes into account the sum of rating points due to variations of four main parameters: soils, unsaturated thickness, net recharge, and river beds.

Trojan and Perry (1988) reported that several other rating systems were used by some state agencies in the United States, like the Hawaii Department of Health (PRZM) and Idaho Department of Health and Welfare (SAFE). Other rating systems developed in the United States, primarily for assessing the sensitivity of aquifers to pesticide contamination were summarized in a U.S. Environmental Protection Agency report (US EPA, 1991).

***Point Count System Models***

A further evolution in the parametric evaluation systems has been the introduction of point count system models (PCSM), also called "parameter weighting and rating methods." They differ from the rating systems that in addition to a rating, a multiplier--identified as an

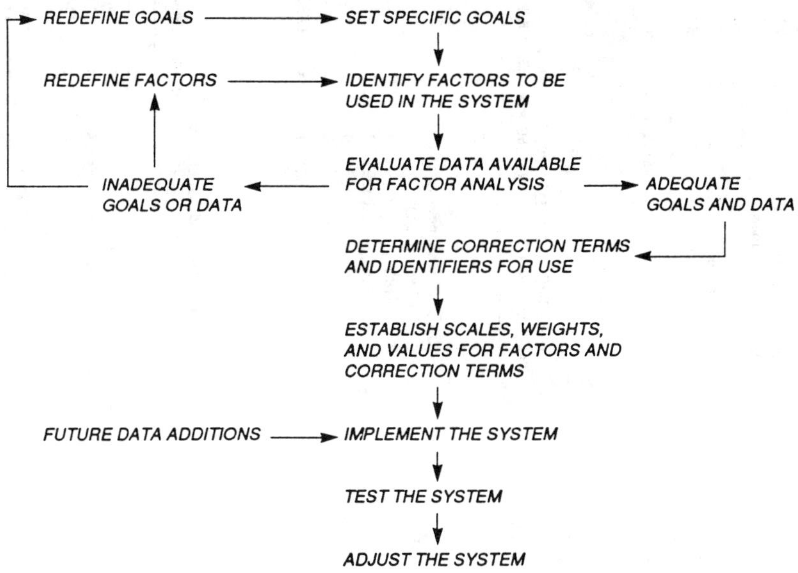


**Figure 9**. Algorithm for developing a parameter weighting and rating system (from Trojan and Perry, 1988).

importance weight--is assigned to each parameter to reflect fairly the relationship among the parameters and their importance for vulnerability assessment. The ratings for each interval are multiplied by the weight for the parameter and the products are summed to obtain the final numerical score that provides relative measure of the vulnerability of one area compared to other areas. The higher the score, the greater sensitivity of an area. Figure 9 shows an algorithm useful in developing a parameter weighing and rating system.

The most difficult aspect of implementing the parameter weighting and rating method is to break the final numerical score range into general classes of vulnerability (for example: highly, moderately, and least vulnerable). Choosing the scores that separate the classes is judgmental and requires a substantial expertise of the map preparer.

One of the first PCSMs was developed for the U.S. Environmental Protection Agency in 1985 by Aller et al (1987), with the acronym DRASTIC (see list of acronyms), taken from initial letters of seven parameters (see Table 6) used to evaluate intrinsic vulnerability of aquifers (i.e., groundwater contamination potential). Each parameter is given a rating interval from 1 to 10, with two relative weight strings (varying from 1 to 5). The most significant parameters have weights of 5; the least significant, a weight of 1. The second weight string was developed to reflect the effect of agricultural activities, in particular, pesticides. In both cases, the index is made up by a sum of products rating for weight of the seven parameters. A computational example is shown on Figure 10.

S1 - Outcropping Gneiss

| FACTOR | D A T A | RATING | * WEIGTH= | NUM. |
|---|---|---|---|---|
| D | >30 m | 1 | 5 | 5 |
| R | 300 mm | 9 | 4 | 36 |
| A | Gneiss, fract. | 5 | 3 | 15 |
| S | Absent | 10 | 2 | 20 |
| T | >18% | 1 | 1 | 1 |
| I | Gneiss, fract. | 4 | 5 | 20 |
| C | E-05 m/s | 9 | 3 | 27 |
| | DRASTIC Index | | | 124 |

S2 - Glacial deposits over Gneiss

| FACTOR | D A T A | RATING | * WEIGHT= | NUM. |
|---|---|---|---|---|
| D | 25 m | 2 | 5 | 10 |
| R | 300 mm | 9 | 4 | 36 |
| A | Gneiss, fract. | 5 | 3 | 15 |
| S | Sandy | 9 | 2 | 18 |
| T | 13% | 3 | 1 | 3 |
| I | Sand/grav/gneiss | 5 | 5 | 25 |
| C | E-05 m/s | 9 | 3 | 27 |
| | DRASTIC Index | | | 134 |

S3 - Alluvial deposits

| FACTOR | D A T A | RATING | * WEIGHT= | NUM. |
|---|---|---|---|---|
| D | 3 m | 10 | 5 | 50 |
| R | 300 mm | 9 | 4 | 36 |
| A | Gravel & Sand | 8 | 3 | 24 |
| S | Sandy loam | 6 | 2 | 12 |
| T | 2% | 10 | 1 | 10 |
| I | Gravel & Sand | 8 | 5 | 40 |
| C | E-02 m/s | 10 | 3 | 30 |
| | DRASTIC Index | | | 202 |

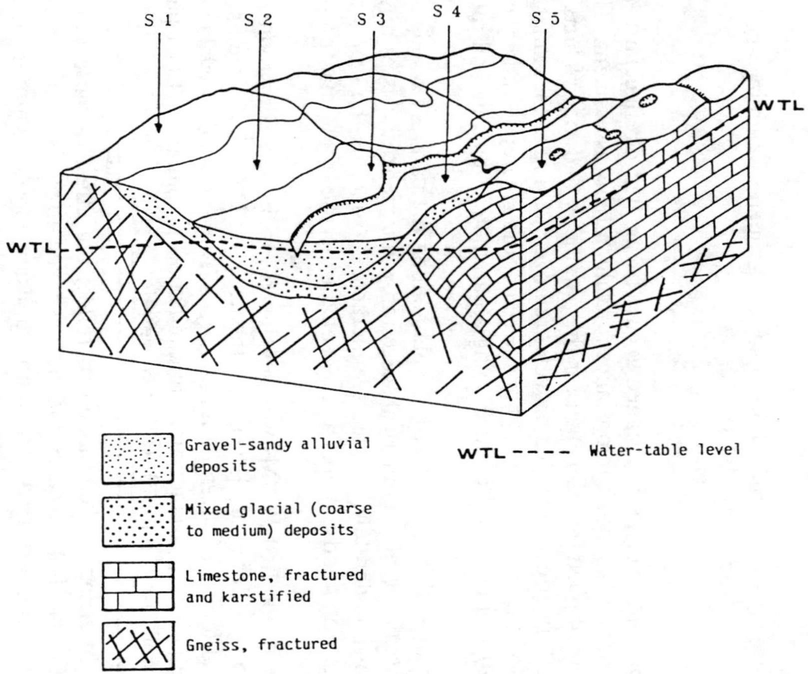


S4 - Glacial deposit over Limestone

| FACTOR | D A T A | RATING | * WEIGHT= | NUM. |
|---|---|---|---|---|
| D | 10 m | 7 | 5 | 35 |
| R | 300 mm | 9 | 4 | 36 |
| A | Limestone, karst. | 9 | 3 | 27 |
| S | Sandy loam | 6 | 2 | 12 |
| T | 18% | 3 | 1 | 3 |
| I | Sand/grav/Limes. | 7 | 5 | 35 |
| C | E-03 m/s | 9 | 3 | 27 |
| | DRASTIC Index | | | 175 |

S5 - Outcropping Limestone

| FACTOR | D A T A | RATING | * WEIGHT= | NUM. |
|---|---|---|---|---|
| D | >30 | 1 | 5 | 5 |
| R | 300 mm | 9 | 4 | 36 |
| A | Limestone, karst. | 9 | 3 | 27 |
| S | Absent | 10 | 2 | 20 |
| T | >18% | 1 | 1 | 1 |
| I | Limestone karst. | 10 | 5 | 50 |
| C | E-03 m/s | 9 | 3 | 27 |
| | DRASTIC Index | | | 166 |

**Figure 10.** Examples of DRASTIC index in various hydrogeological settings (from Civita 1990a).

Once a DRASTIC index has been computed, it is possible to identify areas which are more likely to be susceptible to groundwater contamination relative to one another. The higher the DRASTIC index, the greater the groundwater contamination potential. The DRASTIC index provides only a relative evaluation tool and is not designed to provide absolute answers. Therefore, the numbers generated in the DRASTIC index and in the Pesticide DRASTIC index cannot be equated.

Since its inception in 1985 and testing on 11 county demonstration maps across the United States, DRASTIC has been used by many agencies--sometimes effectively, sometimes not. Its effectiveness has still to be proven because of its limitations. Its main weakness seems to be that it is not flexible enough to be customized to specific needs. So many variables are factored into the final number (vulnerability index) that critical parameters in the groundwater vulnerability may be subdued by other parameters that have no bearing on vulnerability for a particular setting.

Several users of DRASTIC identified a number of shortcomings with this system and tried to deal with these difficulties by adjustments and modifications (Cavallin et al, 1987; Evans and Myers, 1990; Lance et al, 1991; Liddle et al, 1989; Moore, 1988; Rosen, 1994; Trojan and Perry, 1988; US EPA, 1991; Zaporozec, 1987).

In Italy research started in 1990 to develop a PCSM derived from DRASTIC experience, but properly corrected and adjusted to overcome the problems mentioned above and to have a methodology better suited for vulnerability assessment and mapping at a medium to small scale as required in the Italian highly diversified hydrogeology (Civita in Benacchio et al, 1988; Civita, 1990a). The system, provisionally named SINTACS (see list of acronyms), has a complex structure (Figure 11a and b). It is entirely computerized, both for the discretized input stage (grid square) and for the output (mapping and numerical tables). The input data may be coded according to the real situation in the tested area. A number of weight strings, in parallel and not in series, are used to define the effective condition of possible impact. The relative weights of parameters used in SINTACS are shown in Figure 12. The resulting indexes are percentized, divided into intervals that have been defined on the basis of some 500 tests, and grouped into six vulnerability classes. The system has already been tested on two test sites; one in a large plain southward of Torino (Civita, Chiappone et al, 1990), and one in the karstic massif of the Apuanian Alps (Civita, Forti et al, 1990).

Several other PCSMs have been recently presented, the most interesting of which was proposed by Trojan and Perry (1988). A hazard index representing the "hydrogeologic sensitivity" of a region is computed using weights and scores of a variable number of parameters (see Table 6), adjusted and/or integrated by "identifiers" and "correctors" to fit the method to each setting as well as possible.

Vulnerability assessment and mapping should be primarily based on hydrogeological evaluation, rather than on general, automatic rating procedures. A combination of aquifer simulation models and geographical information systems offers a unique opportunity to perform this task.

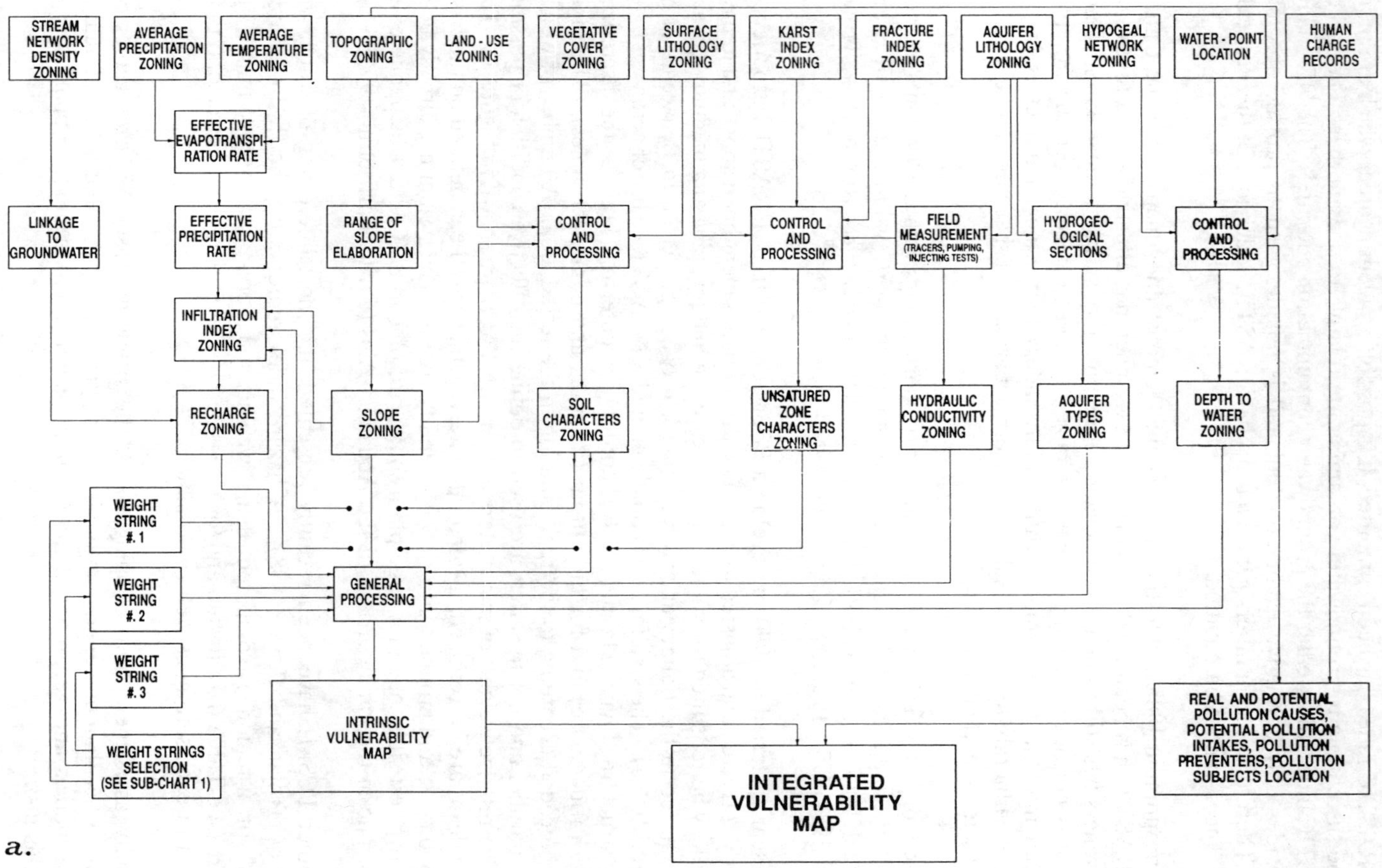


*a.*

**Figure 11.** The SINTACS parametric system (from Civita, Forti et al, 1991): a - main flow chart; b - subchart (p.47).

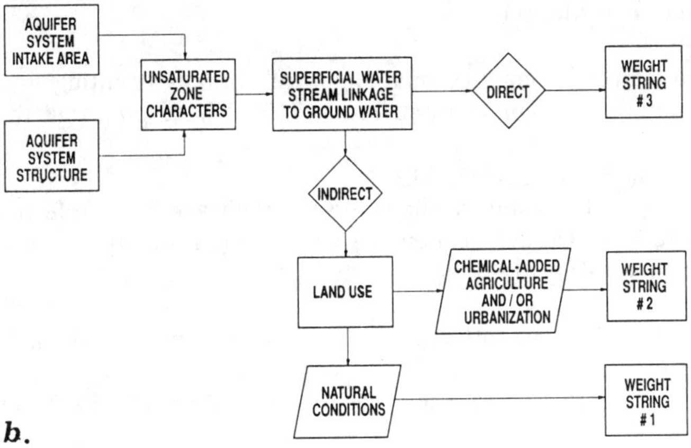


*b.*

**Figure 11b.** Weight strings selection for SINTACS.

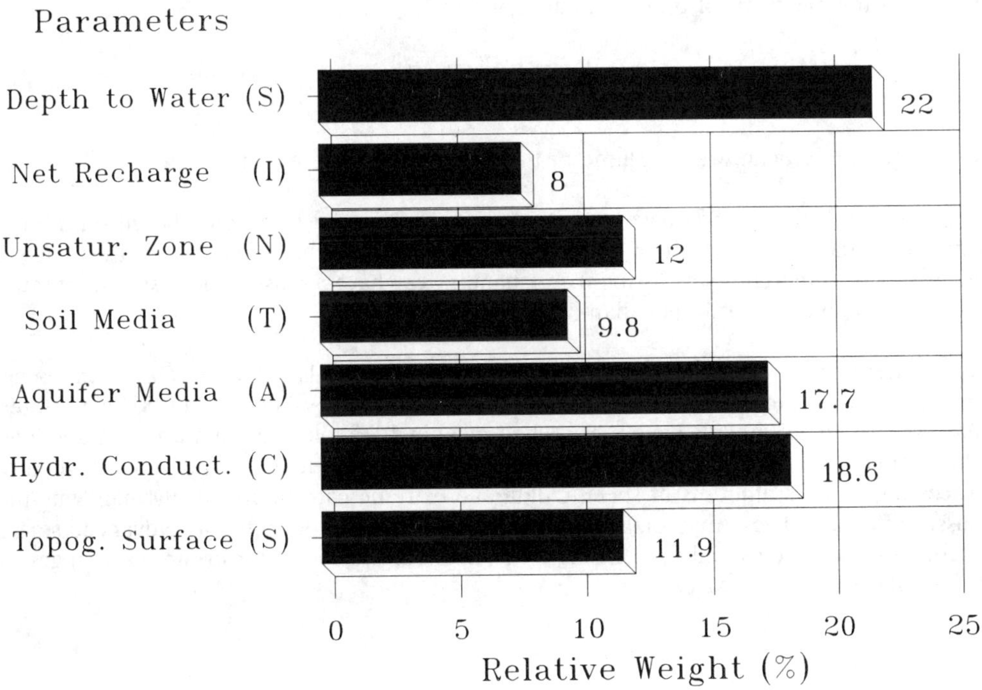


**Figure 12.** Relative weights of SINTACS parameters in intrinsic vulnerability assessment of the La Loggia-Carignano area (from Civita, Chiappone et al, 1990).

**Analalogical Relations and Numerical Models**

These techniques are based on simple or complex mathematical symbols resulting in a vulnerability index ($I_v$). For example, Marcolongo and Pretto (1987) have proposed the Darcy-derived expression:

$$I_V = [K\ (QI/SI)]/MS$$

which actually gives an evaluation of vulnerability as the inverse of the travel time, referred to as a "piston-flow" model; where K = hydraulic conductivity of the unsaturated thickness (SI), MS = actual soil moisture, and QI = the infiltration rate per unit surface.

A similar approach has been used by the Committee of European Economic Communities (EEC) for a large-scale mapping in the second phase of a research program on groundwater resources of EEC (Fried, 1987; Zampetti, 1983). A test was made by Meinardi (1982) for the territory of Holland.

An interesting technique, although still untested, has been proposed by Andersen and Gosk (1987). They included in vulnerability evaluation only two factors, namely the cleansing capacity of soils and the restoration capability of aquifer. The first factor is to be evaluated case by case, as a function of soil type and of contaminants. It should be expressed as a contaminant quantity removed by the unit volume of soil. The second factor (Cr) is the inverse of the mean travel time in an aquifer (Tr):

$$Cr = I/Vw \text{ years}^{-1}$$
$$Tr = Vw/I \text{ years}$$

where Vw is the mean water volume of the aquifer and I the effective yearly infiltration.

Bachmat and Collin (1987) have proposed a complex model technique, based on a large number of data, most of them difficult to gather. It is questionable whether such a technique for vulnerability assessment and mapping can have a reasonable cost-benefit ratio, even for a land area not larger than a hectare.

Also questionable is the applicability of numerical models in vulnerability mapping. Difficulties associated with this method were best expressed by LeGrand (1983) who stated that "Mathematical models have merit when meaningful geological data are available and where there are historical records of contaminant movement. The models require processing of large amounts of specific data and extreme care in effectively managing the masses of data... These mathematical approaches appear to be suited for advanced stages of contaminant studies and for long-term formal studies, but not for preliminary stages or places where data are scarce."

Chapter 6.
# GROUNDWATER VULNERABILITY IN AREAS OF CLIMATIC EXTREMES

Previous chapters have addressed groundwater vulnerability in temperate climatic conditions under normal temperature and precipitation. However, the situation changes somewhat as climatic extremes are approached. Both the conditions of the geological materials and the rate of flow and nature of the groundwater and contaminants flowing through them, and the reactions of these liquids and solids with each other, can be different than encountered in temperate climates.

A literature search revealed that there was, generally, minimal consideration of the topic of aquifer vulnerability under climatic extremes. The obvious reason being that most areas where groundwater contamination has been studied are in temperate regions; most researchers in those areas have, hence, focused their efforts on aquifer vulnerability in temperate climates. This chapter provides a brief overview of aquifer vulnerability under conditions of climatic extremes, focusing largely on arid zones.

## CONDITIONS INFLUENCING VULNERABILITY

There are at least four conditions that influence groundwater vulnerability in regions of climatic extremes: extreme dryness, extreme precipitation, extreme heat, and extreme cold. Each of these has its own domain of impact.

For example, substances such as organic compounds have specific physical characteristics affecting their mobility. An example of this might be volatility. In hot climates, a large portion of an organic substance released as a surface spill might be rapidly released to the atmosphere, never entering groundwater. In a cold climate, the contaminant would remain at the surface for a greater period of time and would more likely enter the groundwater flow system. Solubility also plays a major role, with contaminant solubility in water generally increasing with temperature. In a wet environment, contaminants would tend to quickly enter into solution and groundwater flow. However, extremely wet conditions would favor overland flow of such contaminants to surface water rather than as recharge to groundwater.

Also, the role of individual parameters varies with the change of climatic conditions. For example, the distance to the water table would have a more important role in contaminant attenuation in dry climates than in wet climates. Frequently, the greater depth to the water table in dry regions favors a lessening of contaminant impact on groundwater resources. However, very rapid recharge rates may result in little more than dilution; sufficient contact time is needed between contaminants and the earth materials of the unsaturated zone to result in meaningful attenuation.

## EXTREME DRYNESS

Arid regions are noted for very low rates of recharge to aquifers because the amount of potential evaporation greatly exceeds the rate of precipitation. If contaminants are released at or below ground surface, there is a very slow downward movement of contaminants. Thus, if the contamination event is unknown, the result is delayed detection. In fact, considering this very slow movement of contaminants, the actual contamination event may have occurred many years earlier.

At the same time, there is minimal dilution of contaminants. Yet, there is more soil-rock contact time for contaminants borne in percolating groundwater, a favorable situation. Additionally, arid regions can have naturally cemented surface horizons (e.g., the *caliche* soils of northern Mexico), which may prevent rapid movement downward, and allow considerably more time for both the physical and chemical processes and regulatory agencies to react to spills and releases of contaminants.

Handa (1983) noted that concentrations of nitrate and phosphate in the groundwaters of arid or semiarid regions of India were higher than in the humid regions of eastern and southwestern India. The difference is attributed to the anaerobic soil conditions of humid flooded lands as compared to the dry areas where better drainage occurs.

In the northern part of Mexico, diluted industrial solvents have moved downward through fractured calcareous siltstones and fine-grained sandstones (J. Miller, personal communication, 1990). Although the annual rainfall is low (approximately 340 mm), with a minimal surcharge of precipitation over evaporation, such contaminants, discharged into unlined holding ponds, readily moved to the water table at a depth of 10 m. Under normal conditions of a low rate of precipitation, the contaminants may not have reached the water table; however, the hydraulic head difference established by the wastewater ponds and the on-site pumping wells resulted in a considerably more rapid rate of flow. Normally, groundwater recharge in this region occurs only during periods of heavy hurricane-related rainfall, with the actual recharge being from short-lived streams flowing around the site.

### Sensitivity of Aquifers to Natural Impacts

Water and energy cycles in arid zones take on special characteristics because of the deficient and variable rainfall, abundant solar energy, and cloudless skies. Occasionally, rainfall will be sudden and heavy with associated flash flooding, but it tends to be lost rapidly through evaporation. Dry lands are sensitive to minor shifts in their water and energy balances. They tend to encompass climatic belts that are progressively more arid inward. Variance in rainfall increases as aridity increases and, therefore, areas most subject to drought are those in which the variations in annual rainfall are greatest (Rassam, 1988). However, since drought (meteorological, hydrological, or agricultural) originates from a deficiency in precipitation, it results in a water shortage for some human activities. Endeavors that are dependent on rainfall are virtually nonexistent in desert and hyperarid desert zones. Serious water problems are encountered in semiarid zones (Bakour and Kolars, 1994; Ibrahim, 1993).

Figure 13 shows how the variations in precipitation relate to aridity. The zone of greatest unpredictability lies at the intersection of the two curves; this occurs in semiarid zones. These climatic zones are the most sensitive to drought.

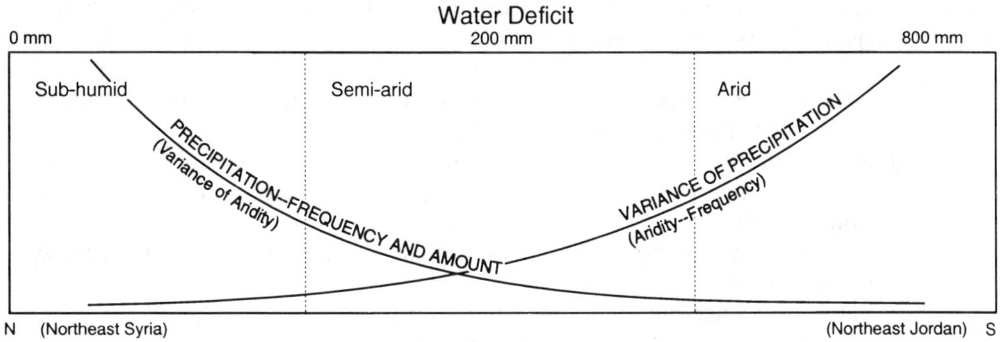


**Figure 13.** Variance of precipitation and aridity in arid and semiarid zones in the eastern Mediterranean region (after Bakour and Kolars, 1994).

Vulnerability has been previously defined in this book as "an intrinsic property of a groundwater system that depends on the sensitivity of that system to human and/or natural impacts" (Chapter 2). This definition can include quantity as well as quality of groundwater. The development of the concept of groundwater vulnerability to include quantitative aspects may stimulate mapping activities in this context. Such maps would be of considerable value for the estimate of potential vulnerability of groundwater to drought, and would be an important tool for drought preparedness planning. This type of map would insure a suitable response to water shortages during future droughts. Groundwater, being a major source for water supplies during drought periods, could be drawn upon without inflicting irreparable damage to the resource.

The sensitivity of an aquifer to drought depends on the amount and mode of recharge, which could be in the form of direct recharge from precipitation or of indirect recharge from "wadi" flow. The latter is the most important source of recharge in semiarid zones (Edmunds et al, 1987). Using geochemical techniques, these researchers estimated direct annual recharge to the Nubian sandstone regional aquifer system in Sudan to be only 1 mm (mean annual rainfall is about 200 mm).

It would be equally pertinent to consider groundwater vulnerability to desertification, since desertification tends to increase runoff and decrease infiltration. Areas of aquifers that are sensitive to drought are usually the most vulnerable to desertification.

**Sensitivity of Aquifers to Human Impacts**

A review and analysis of the impact of intensive developments on groundwater resources and the environment in arid and semiarid zones (Ibrahim, 1993; Khouri, 1993; Llamas et al, 1992) revealed that existing and past development (or overdevelopment) has resulted in several adverse impacts. On the basis of case studies (Ibrahim, 1993; Khouri 1993), the following negative impacts have been identified:

(a) Depletion of aquifers has become a serious problem in several regions (e.g., Arabian Peninsula and North Africa).
(b) Rapid and excessive decline in water levels have economic implications and are related to aquifer depletion, when a significant part of groundwater is taken from storage.
(c) Deterioration of groundwater quality, due to sea water intrusion or to up/downconing from underlying or overlying saline water bodies.
(d) Land subsidence due to groundwater extraction.
(e) Salinization of the soil and shallow aquifer systems.
(f) Contamination of shallow groundwater.

Llamas et al (1992) investigated the same phenomena on a global scale and concluded that, in addition to the above-mentioned effects, environmental impacts on aquatic systems have occurred in arid and semiarid zones.

The assessment of groundwater vulnerability to contamination, salinization, and depletion could assist planners in minimizing adverse effects of groundwater development. Vulnerability needs to be assessed in arid zones always in terms of quantity and quality, because it is often difficult in these regions to separate quantitative and qualitative aspects.

The introduction of irrigation disturbs the delicate water balance in arid lands. The return flow of irrigation water is normally saltier than the water in the underlying aquifer system. The salinity of groundwater would tend to increase slowly but steadily, because return flow in arid climates constitutes a significant portion of recharge (Llamas et al, 1992). Moreover, the soil may contain cemented horizons or "hardpans" that slow down infiltration of irrigation water; then, evaporation will last longer and the water gradually becomes saltier. The degree of salinization depends upon the thickness and depth of these horizons.

Contamination of aquifer systems is a serious problem in the urbanized and irrigated belts of arid lands. An assessment of aquifer vulnerability and the preparation of specific vulnerability maps to evaluate the impact of a particular land use on groundwater systems in arid and semiarid zones is an issue of fundamental importance for future planning, protection, and management of these vulnerable groundwater resources.

**Assessment of Natural Vulnerability**

Recharge, assessed usually as net recharge, is an important attribute for the assessment of the vulnerability of groundwater systems to drought in subhumid and in semiarid zones (regions that receive annual precipitation ranging from 200 to 500 mm). The sensitivity to drought episodes increases with increasing aridity (see Figure 13). In arid and hyperarid regions ($<100$ mm per year), recharge is negligible, and the aquifers which now occur in such environments were recharged during the humid episodes of the Quaternary Period (Khouri, 1989). They are, therefore, independent of present climatic conditions, and are, of course, not vulnerable to drought.

The ratio of the rate of recharge to the volume of water in storage is considered more appropriate for the assessment of vulnerability of groundwater in arid and semiarid zones than net recharge alone. It is particularly significant for the evaluation of vulnerability to drought. Andersen and Gosk (1987) used the term "restoration capacity" of the aquifer for the ratio of the volume of water in storage ($m^3$) to the rate of recharge ($m^3$/year).

Soil is usually poorly developed in the greater part of arid and semiarid lands. It should, therefore, be considered with the unsaturated zone as one physical unit. In agricultural lands, salts often accumulate in soils. Soils in such areas have negative impacts, and do not act, as they normally do, as protective or purifying media. It is, therefore, imperative in assessing the influence of this attribute on vulnerability to differentiate between soils under human stress from soils under natural conditions.

Soils in arid zones are highly vulnerable, and are subject to desertification. Dry land soils stressed by drought and disturbed by land use will be exposed to wind and water erosion. The exposed surfaces of silty and clayey soils are hardened in such zones. The process of surface sealing leads to one of the most detrimental, long-lasting effects of overuse of land, reducing infiltration and increasing runoff. Where the soil is deficient in moisture, the moisture demand must be satisfied before water can penetrate to the water table.

Unconfined aquifers are more sensitive to drought than confined aquifers. Aquifers of limited thickness and areal extent, low storage capacity, and low hydraulic conductivity are the most vulnerable to drought (Vrba, 1991).

The response of aquifer systems to development, which is a structural characteristic, could be much more significant than the aquifer sensitivity to variable recharge. Such structural sensitivity, if combined with the risk to drought that affects the groundwater system and its "resistance" to drought, could be assessed, classified, and mapped on a small scale (Vrba, 1991). It is a useful concept for planning water supply projects, since it indicates the reliability of the resource.

The response of a groundwater reservoir to drought may range from immediate to very slow depending on the nature and the extent of the aquifer, depth to water, and, for a confined aquifer, also on the extent and distance of recharge areas. Therefore, the full impact of drought on groundwater systems may not be immediately apparent. This response does not affect, however, assessment results.

In fact, water levels, and particularly artesian pressures, show little decline early in a drought, but they may continue to decline for sometime following the end of the drought. By lagging in their response to meteorological events, groundwater systems have a stabilizing influence on streamflow. Extensive and large groundwater reservoirs provide a tremendous reserve of usable water in addition to playing a stabilizing role in the functioning of the entire hydrologic system.

**Assessment of Specific Vulnerability**

The hydrogeological characteristics of aquifers, land use practices, and the contaminant characteristics and loading all have to be considered in assessing the specific vulnerability (see Chapter 5).

The second half of the 20th century has witnessed a rapid growth of urban areas in arid zones for a variety of reasons, among them economics and tourism. Underlying aquifer systems and aquifers in neighboring areas have been intensively developed to meet rising demands (Khouri, 1993; Llamas et al, 1992). Aquifer vulnerability to both depletion and contamination need to be assessed since these two factors are often interconnected. Total recharge should be considered. It has been observed that a rapid rise in the water table has occurred in the majority of urban centers. This phenomenon is attributed to leakage from water supply and sanitation networks combined with the presence of hardpans and impermeable layers in the unsaturated zone.

Besides contamination, major human impacts in arid climates include aquifer depletion and salinization. Custodio (1990) suggested the use of isotopic techniques for the investigation of aquifer vulnerability to salinization and contamination. The principal factors that need to be investigated for the assessment of specific groundwater vulnerability in arid lands are land use and population density. Normally, there is a sharp contrast between areas under stress (irrigation and urbanization) and areas where natural conditions prevail. The major parts of arid zones are sparsely populated; deserts and hyperdeserts are virtually uninhabited areas. They are usually underlain by deep regional aquifer systems, which essentially are nonrenewable. These factors lower the vulnerability to almost nil (Margat, 1992). By contrast, land irrigated by surface water (in large or small river basins) or by groundwater (in oases) are usually under severe stress.

Assessment of vulnerability of groundwater to salinization may require a special approach and an understanding of the conditions that cause salinization. The origin of salts varies widely. They may be derived from parent rocks that release salts during the weathering process. Irrigation water always contains some salts, which generally tend to accumulate under high rates of evaporation.

Among the major factors that influence the water and salt balance in the unsaturated zone is the infiltration of irrigation water. Salinization occurs when the amount of salts accumulating is greater than that of salts removed. The salinity of the soil should not exceed acceptable limits for the particular crop grown. This is accomplished by applying irrigation water in excess to that needed for consumptive use. This practice transfers the salinity problem from the soil to the groundwater. In the early stages of the buildup of groundwater levels, deep percolation leaches the most soluble constituents to depth, particularly the more mobile constituents such as chlorides. When and where the water table is at or above a "critical depth", capillary movement from the water table to the soil surface will take place. The critical depth varies with the physical characteristics of soils (primarily light or heavy texture).

The water table, however, does not control the level of salinity or the rate of salinization, but the direction of the flux does (Van Schilfgaarde, 1984). Where water and salt have moved upwards into the soil profile, one expects to find an inverted salinity profile (salinity decreases with depth). Where the dominant movement is downward, a normal salinity profile is usually encountered (salinity increases with depth). The depth of the water table is a useful diagnostic parameter in assessing the salinity hazard. Assessment of vulnerability to salinization thus entails forecasting of the expected changes in the elevation of the water table under conditions of irrigation, predictable by using simulation techniques. Other methods have been proposed (Peezely, 1976), based on the characterization of the natural water balance of groundwater and the changes caused by irrigation and, finally, on the characterization of the new equilibrium.

To summarize, the main attributes that need to be assessed in arid regions are recharge (from irrigation return flow, seepage from irrigation networks and natural streams); soil properties, particularly the presence and absence of hardpans; aquifer characteristics, particularly the hydraulic conductivity; hydraulic gradient; and topography.

## EXTREME WETNESS

Regions receiving a much larger quantity of rainfall potentially experience greater recharge; although surface runoff is very high. Rapid downward movement of contaminants can occur, followed by possible rapid lateral movement with groundwater flow. A high rate of dilution of contaminants might be expected, depending upon the ability of the contaminants to mix with the groundwater. At the same time, there is a lesser soil-rock contact time for contaminants borne in percolating groundwater. And, there is little time to react to spill events.

It is estimated that the combination of heavy rainfall (1800 mm) and highly permeable volcanic ash in the Hawaiian Islands, USA results in a 30-percent recharge rate (Aller et al, 1987). Such a recharge rate, while it may favor attenuation through dilution, also allows for very rapid lateral movement of contaminants and minimal response time to contaminant spill incidents.

## EXTREME HEAT

Hot climates can be either dry or wet, often with abrupt seasonal changes. Heat is not a controlling factor, except that the contaminants and associated groundwater will be warmer and may be more reactive with the soil/rock materials. However, such substances will be more mobile. It is also to be expected that soil microorganisms will be more abundant and more active, favoring breakdown of contaminants, specifically organic substances.

## EXTREME COLD

Frozen soils tend to inhibit the downward movement of contaminants. The flow path of groundwater through permafrost is complicated and difficult to predict.

Pinneker (1974) discussed the protection of the groundwaters of Siberia, a phenomenon that must take into account both aridity and low temperature. He pointed out that the "self-cleaning" ability of the groundwater is considerably lower in the permafrost regions than in unfrozen ones. The freeze/thaw process results in an increase in mineralization of the groundwater; sewage contaminants tend to concentrate rather than becoming diluted. This process also appears to be irreversible.

Despite the cold temperatures of Alaska, USA, groundwater microorganisms are, apparently, abundant and reactive with contaminants there. In areas with solvent- and fuel-related spills or leaks, it was noted that the groundwaters contained a high level of alkalinity, due to the release of carbon dioxide by active fuel-eating bacteria (J. Miller, personal communication, 1991). That is an indication that beneficial bacteria are found even at low groundwater temperatures.

## CONCLUDING REMARKS

Considering the present minimal knowledge of aquifer vulnerability in regions of climatic extremes, hydrogeologists should maintain an awareness of these conditions. Vulnerability as perceived in temperate climates can be considerable greater or less. Building upon a basic knowledge of the hydrogeological framework and extremes of the climatic conditions, hydrogeologists should consider vulnerability from a less-than-straightforward approach. And, the reactions of potential contaminants under such conditions to the hydrogeological system should also be taken into account.

Chapter 7.

# DATA NEEDS AND PRESENTATION

## DATA NEEDS AND ACQUISITION

### Data Needs

Vulnerability assessment requires a thorough knowledge of local hydrogeological, hydrochemical, and contamination data and of the type and location of potential contamination sources. Collection of such data can be quite difficult and expensive. So, it is imperative to use all available information and to spend as little effort as possible in acquiring new data.

Table 7 shows basic information needed for groundwater vulnerability assessment and mapping; the technical and scientific organizations and sources commonly supplying needed information; and direct or indirect methods to be used to collect and/or supplement existing information.

### Information Sources and Data Collection Methods

Hydrogeological, hydrological, and environmental information may be gathered in different ways depending on the extent of an area and its morphological, climatological, and land-use conditions. In developed countries much data may be supplied by public agencies, universities, scientific institutions, geological or/and resource exploration companies, statistical and census authorities, consulting firms, etc. (Table 7). Unfortunately, the same type and number of facilities are not as available in the less developed countries and in poorly developed or sparsely inhabited areas. In these areas it is necessary to start from the beginning in order to acquire a complete but quite expensive data base.

The methods and techniques for data collection and processing are common practices for people working in the investigation and protection of groundwater resources. Therefore, discussion of the methods of conducting field surveys and making hydrological measurements, or of procedures for performing special water tests and analyses was not included in this book. However, it is useful to emphasize the application of unconventional techniques, such as remote sensing, that seem to be promising tools in natural resource exploration and environmental monitoring and control.

### Remote Sensing

Some of the widely used remote sensing techniques for the collection of data needed for an assessment of groundwater vulnerability and for the construction of vulnerability maps include:

**Table 7.** Information base components and potential sources of data needed for the assessment and mapping of groundwater vulnerability (modified from Civita 1993).

| BASIC RESOURCE FIELD | INFORMATION TYPE | DATA SOURCE | | | | | | | | | | FIELD DATA | | | |
|---|---|---|---|---|---|---|---|---|---|---|---|---|---|---|---|
| | | AGRICULTURE AGENCIES & FARMER ORGANIZATIONS | WATER SUPPLY AGENCIES AUTHORITIES & COMPANIES | PUBLIC HEALTH & ENVIRON-MENTAL PROTECTION AUTH. | EARTH SCIENCE RESEARCH CENTRES & UNIVERSITIES | CONSULTING FIRMS, COMP. & PROFESSIONAL ASSOC. | CLIMATOLOGICAL & HYDRO-LOGICAL DATA RECORDS | TERRITORIAL PLANNING AUTHORITIES & DEPARTM. | DRILLING GEOPHYSICS & GEOTECHNICAL COMPANIES | STATE & PROVINCIAL GEOLOGICAL SURVEYS | SPECIFIC THEMATIC MAPPING & DATA-BASE | FIELD SURVEY, GAUGING & TESTING OPERATIONS | MONITORING WELLS WORK DATA | AERIAL PHOTO & RE-NOTE SENSING SURVEY | CHEMICAL & BIOLOGICAL SAMPLING & ANALYSIS |
| TOPOGRAPHY | Elevation, slope variability of land surface; surface runoff paths, stream network density. | | | | | | | | | ● | ● | | | ● | |
| VEGETATIVE COVER | Land use, subsurface water pathways, recharge and discharge areas, fracture traces and lineaments, contaminant potential. | ● | | | ● | | | ● | | | ● | ● | | ● | |
| CLIMATOLOGY | Long records of precipitation, average temperature, humidity, solar radiation, evaporation, evapotranspiration; effective precipitation assessment. | | | | | | ● | ● | | ● | ● | ● | | | |
| SOILS | Thickness, structure, texture, mineralogy, chemical and physical properties, porosity, permeability, moisture, infiltration capacity. | ● | | | ● | ● | | ● | | ● | ● | ● | | | |
| HYDROLOGY | Streamflow discharge, hydrograph analysis, baseflow, flow ratio, water exchanges with underlying ground water systems. | ● | | | ● | ● | ● | ● | | | ● | ● | | ● | |
| HYDROGEOLOGY | | | | | | | | | | | | | | | |
| Unsaturated Zone | Depth to water; thickness, lithostratigraphy, mineralogy, geometry, fracture index, karst index, effective porosity, and saturation ratio of surficial deposits; vertical effective permeability, effective flow velocity, infiltration rate index, net recharge. | ● | ● | | ● | ● | | ● | ● | ● | ● | ● | ● | ● | |

**Table 7.** (continued)

| | | | | | | | | | | | | | | | |
|---|---|---|---|---|---|---|---|---|---|---|---|---|---|---|---|
| Saturated Zone | Lithostratigraphy, geological structure, geometry, effective porosity, permeability type (primary or secondary), transmissivity, storativity, and hydraulic conductivity of an aquifer; aquifer type (unconfined, semiconfined, confined); water level fluctuations, hydraulic gradient, flow directions, effective flow velocity and discharge, ground water divides, exchanges with surface water bodies or/and adjacent aquifers. | | ● | | ● | ● | | ● | ● | ● | ● | ● | ● | | |
| WATER USE | Water-discharge points (springs, wells) and location of ground water extraction works; surface and ground water sources, distribution, and usage; yield and drawdown of pumping/dewatering plants, location and inflow rate of recharge systems. | ● | ● | | ● | ● | | ● | ● | | ● | ● | ● | ● | |
| CHEMISTRY | | | | | | | | | | | | | | | |
| Hydrochemistry | Physical and chemical properties of surface and ground water, chemical markers, isotope content, age and residence time of water, characteristic ratios; natural surface and ground water quality distribution. | ● | ● | ● | ● | | | ● | ● | ● | ● | ● | ● | | ● |
| Contaminant Features | Changes in water quality; contaminants present and their physical and chemical characteristics, concentration, half-life, persistence, mobility, dispersivity, cation exchange capacity, biodegradability, etc. | | ● | ● | | | | | | ● | | | ● | | ● |
| HUMAN IMPACT ON THE ENVIRONMENT | Extent of urban areas, location and type of industrial complexes, existing and potential contamination sources, potential contamination entries, main objects of protection. | ● | ● | ● | | | | ● | ● | ● | ● | ● | | ● | |

- Black and white and natural color aerial photo interpretation.
- False-color aerial photo interpretation.
- Multispectral Linescanner (MSS) and Synthetic Aperture Radar (SAR) survey image processing and interpretation.

Regarding the first two techniques, low altitude and medium-high altitude aerial photographic survey may be employed. Low altitude survey is used to identify and map the results of human activities (land use, drainage and stream network changes, sources of contamination, settlements and infrastructures, environmental changes and misuses). Medium-high altitude survey gives more general information concerning geological structures and boundaries, fracturing and karst features, vegetative cover, etc. Both MSS and SAR surveys are aircraft- and satellite-vectored. They can provide static (i.e., single image) and dynamic (multiple images) information.

The newest remote sensing, satellite-vectored systems (NIMBUS, LANDSAT, SPOT, HCMM, RADARSAT, Space Shuttle) carry high definition devices that give highly magnifiable images (Figure 14). Recent availability of SOYUZ images (multispectral photographic records with pushed space resolution up to 5 m) has further increased the possibility to delineate phenomena or identify objects at a detailed scale of 1:10 000 - 1:25 000. The most recent generation of satellites for terrestrial resources (LANDSAT 5 and SPOT 1,2) are carrying scanners such as *Thematic Mapper* or *High Resolution Visible*. They have a very high spatial resolution (30 m and 20 or 10 m, respectively, according to X-mode or P-mode data recording) and a specific spectral resolution suitable for lithological, pedological, vegetation, and land-use reconnaissance for vulnerability assessment and mapping (Aller et al, 1987; Marcolongo, personal communication, 1991; Marcolongo and Pretto, 1987).

The most valuable types of information that can be obtained by these systems are:

- distribution of high-rate vertical drainage (high permeability, limited or zero overburden);
- location of permanently wet areas (shallow depth to water, intense seepage from surface water bodies to underlying aquifers);
- existing land use, which allows an evaluation of existing or potential contamination sources (fertilizers, agrochemical substances, etc.);
- vegetative cover condition, commonly affected by changes in underlying rock types, water content of soil, and subsoil and unusual soil chemistry changes (stress within a plant population and resultant changes may be recognized by anomalies in color and reflectance and by emission of radiation at wavelengths outside the visible light spectrum);
- variations in soil texture emphasized by integrating spectral analysis (cluster analysis) and morphological analysis (gravelly, sandy, silty, and clayey soils usually have different spectral signature); and
- hydrogeological complexes and identification of their specific characteristics that can be obtained by interpretation of satellite images and related treatment (thermal inertia mapping, product, ratio, and derivative).

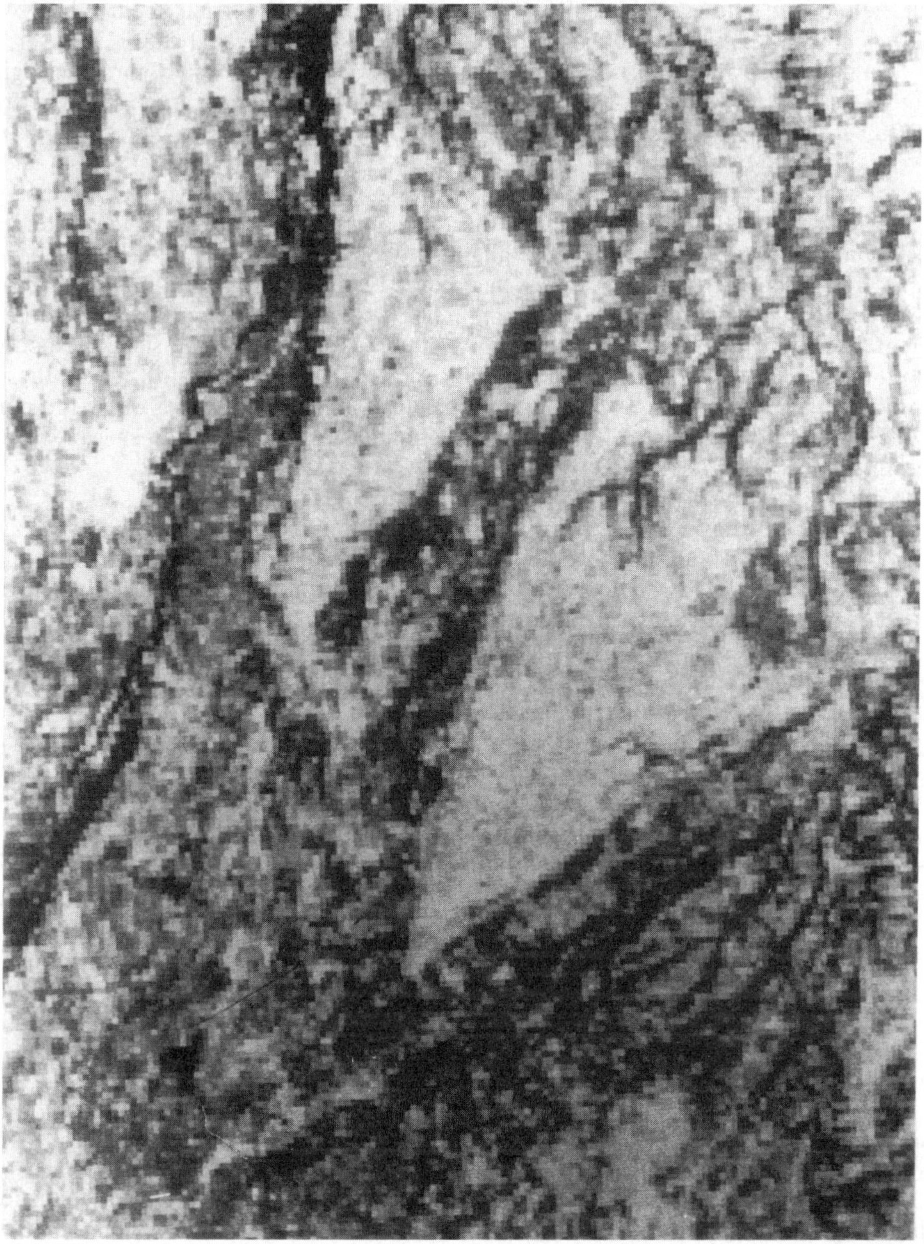

**Figure 14.** LANDSAT 5 thermal infrared band. Light gray zones = sand and gravel terraces (greater depth to water); dark gray = earlier alluvial valley floors (lower depth to water); black = shallow groundwater and discharge to stream network.

In the field of radar imagery, the Shuttle Imaging Radar (SIR) carried aboard the NASA shuttles together with a radiometer (SMIRR) may give good integrated data. However, the best results come from airborne remote sensing. Airborne linescanners provide digital, remotely-sensed data with greater spatial and spectral resolution that can be obtained from LANDSAT or SPOT imagery.

Airborne thermographies in 9-11 $\mu$m band are processed by a harmonic (frequency) analysis that gives a greater detailed description of thermal conditions of the area. Linear elements and discontinuities are indicated by the alignments of thermal gradients and a very high number of discontinuities (much greater than could be obtained by field survey or normal aerial photography) is available for mapping. This is a good basis for a careful statistical study of fracturing (density, prevailing trends, dip directions, etc.). The same method can be used to detect effluent discharges into surface water bodies (Figure 15). Using thermal slicing processing techniques, contaminated effluent (but also fresh water discharge points and/or zones) can be monitored. The flow volume of such discharges can also be determined.

Other image processing, mathematical operations of multiple or single signals (for instance, 1-2 and 9-11 $\mu$m bands) are employed in order to find areas with rapid seepage, subsurface karstic phenomena, interrelationships between groundwater and surface water, and groundwater exchange between adjacent aquifers (Figure 16). Additionally, these procedures can be used to map soil moisture content anomalies due to the presence of shallow groundwater, thermal underground anomalies (e.g., shallow, illegally buried wastes), and soil type and moisture content.

The experience gathered in this manner shows that integration and control of remotely sensed data by even a limited number of ground-control points may give appreciable results for the evaluation of intrinsic vulnerability and for contamination monitoring. Also, remote sensing imagery can be readily merged in a geographical information system (GIS) data base. The greatest limitation to the use of these techniques is that non-systematic errors and distortions are hard to remove, sometimes making accurate mapping difficult. An adequate number of control points, well spaced throughout the area, is essential for digital image correction of multi-band registrations and of thermal and radar scanning.

## PRESENTATION OF DATA ON GROUNDWATER VULNERABILITY MAPS

### Scale and Basic Data

Planning and constructing a vulnerability map involves preliminary evaluation, as realistically as possible, of the number, distribution, and quality of available measurable data. For vulnerability mapping, the best mapping technique will be determined only during the actual evaluation, which will also determine the scale and legend of the map. Considering experience in vulnerability mapping gathered in Italy, it is possible, although only in a qualitative form at the present time, to indicate the correlation between three main factors:

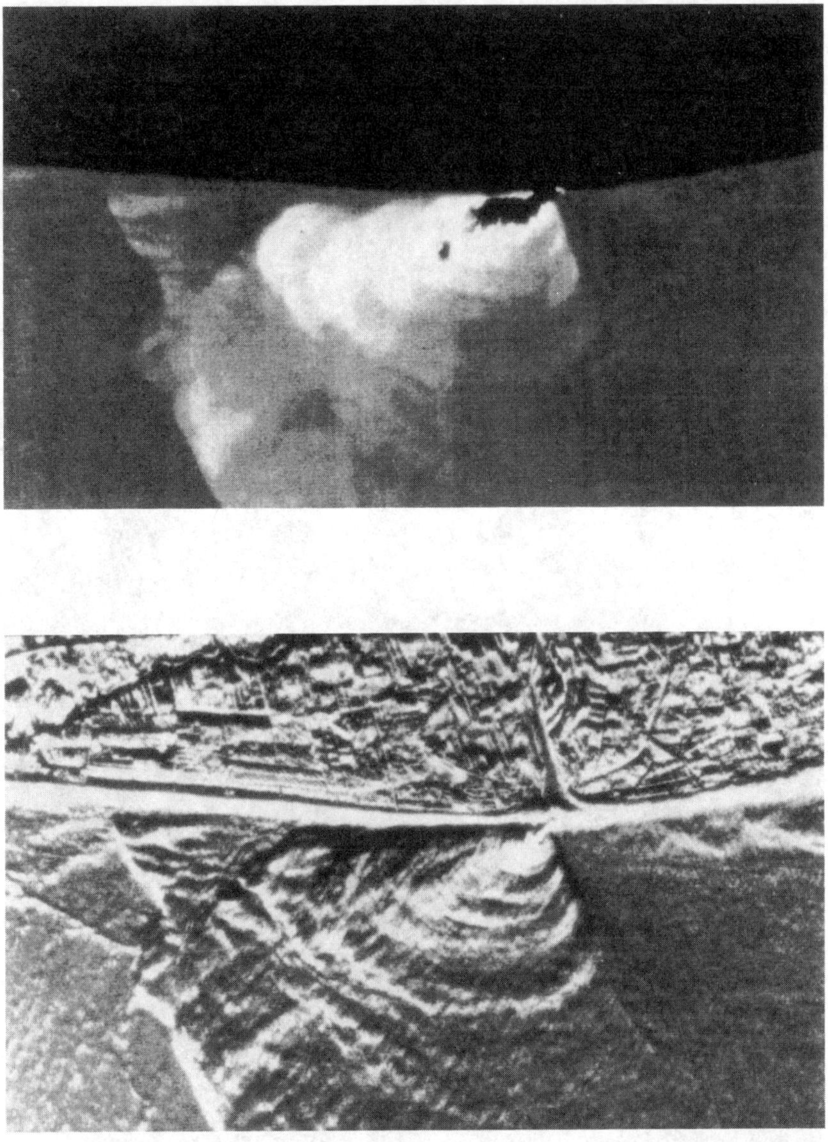

**Figure 15.** Aerial thermal infrared image processing of an effluent discharging into the sea. Top: isothermal levels; bottom: harmonic analysis of 9-11 $\mu$m band (from Civita, Cocozza et al, 1983).

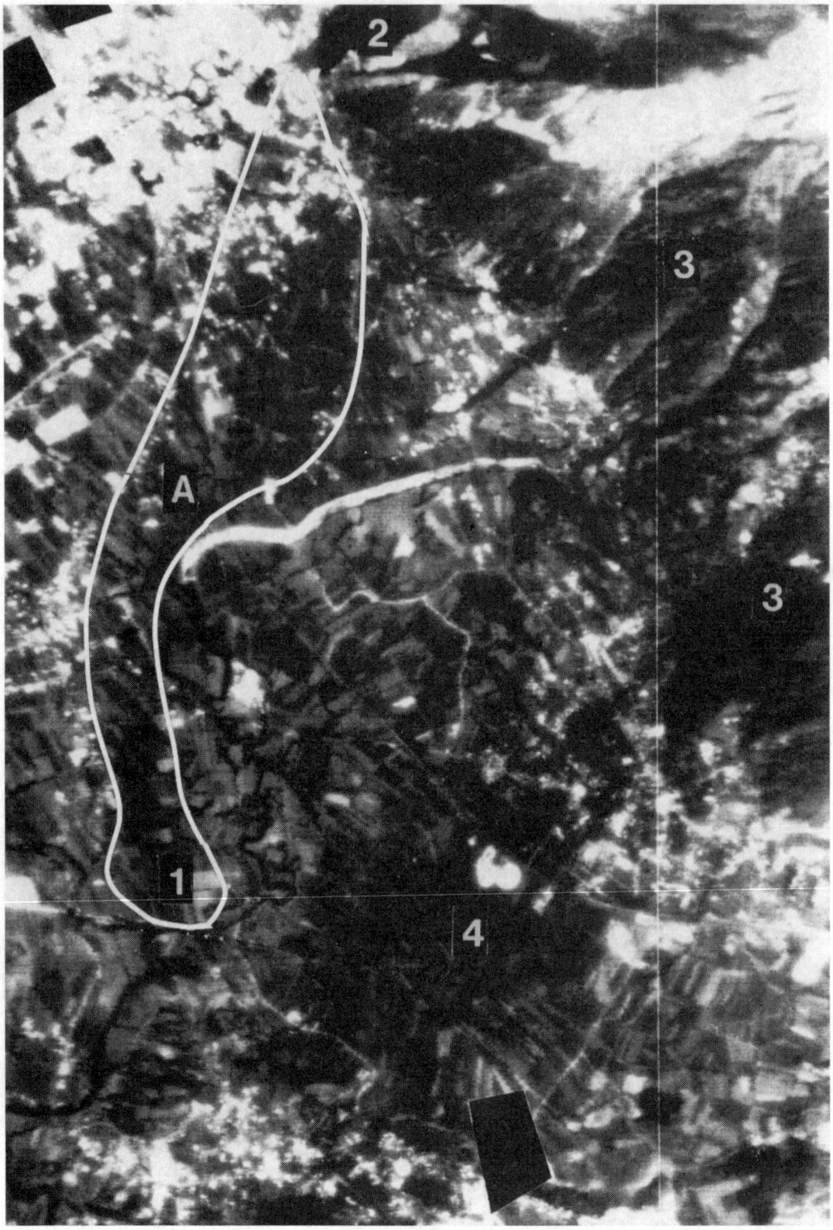


**Figure 16.** Aerial thermal infrared image interpretation. A - groundwater discharge from limestone aquifer to alluvial sand and gravel aquifer; 1 - high-yielding well field (almost 2 $m^3/s$); 2 - limestone (aquifer) mountain side; 3 - flysch (aquiclude) mountain side; 4 - wetland.

(a) the density of surveyed points (IPD), (b) the number of information secured for any point (DNPP), (c) and the scale denominator (SD) at which the map can be constructed. The diagram in Figure 17 shows that:

- complex, low SD models require high density of data points per unit area;
- for medium information data point density having a fair distribution, a more or less complex parametric system (dependent on the number of data available per point) may be used; and
- for areas where density of information points is low and information is scarce and scattered, as is often the case, a hydrogeological complex and setting method with a medium or large SD must be used.

To a large extent, the reliability of basic data is very important consideration in choosing a method for vulnerability assessment. Actually, inadequate data may lead to false precision. Even worse, unreliable data may completely upset results, thus making them useless or misleading. The reliability of data, moreover, can vary widely with the mean elevation of the investigated area. Assigning a range of 1 to 10 to the reliability of data, a variation curve of the data reliability versus the mean elevation can be plotted. As Figure 18 shows, data reliability sharply decreases already at a relatively low altitude (300 to 400 meters above sea level) due to the growing scarcity of available data in mountainous areas, which may only partially be resolved by the use of extrapolation techniques. This is true not only for hydrogeological data (water levels, unsaturated zone, flow directions, hydraulic conductivity, and aquifer geometry), but also for pedological and climatological data (soils, rainfall, evapotranspiration, wind, temperature, etc.).

In mountainous regions and in the majority of hilly areas it may be necessary to avoid the more complex parametric systems and use instead the hydrogeological complex and setting methods or matrix systems, coupled to medium to high SD mapping. The parametric systems have been used more often in flat plains with high data density and reliability, as has been low SD mapping.

**Vulnerability Mapping Approaches**

An intrinsic vulnerability map shows areal changes of a single areal hazard (single hazard, one purpose map). A more sophisticated step (single hazard, multi purpose map) is the specific vulnerability map, which shows the potential of both the soil-rock-aquifer system to contamination and the location of existing and potential contamination sources. For maximum application of this type of map in environmental and groundwater resource planning, the objects in need of protection are also added to maps in order to make the scenario as complete and objective as possible (Civita, 1987a, b; Civita, 1990a).

It is quite evident that some of the basic features of vulnerability maps show no important changes with time (e.g., lithology, geological structure). However, the features linked to human activity are subjected to abrupt and sometimes incidental changes with time. This makes a continous updating of vulnerability maps unavoidable--at least, as far as the human

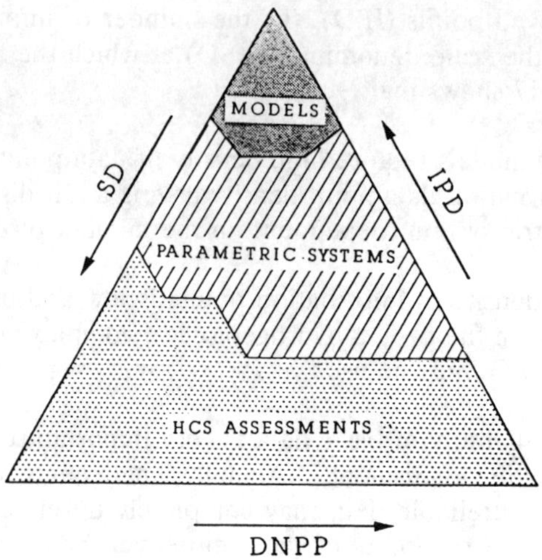


**Figure 17.** Interrelations between map scale denominator (SD), information point density (IPD), and number of data per point (DNPP) for the vulnerability assessment method selection (from Civita 1990a).

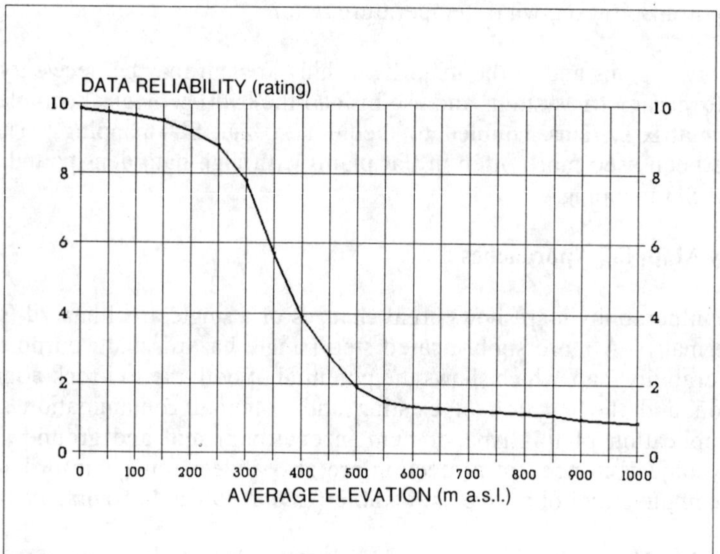


**Figure 18.** Attempt to depict the relation between the basic data reliability and average elevation of an area where an aquifer vulnerability assessment will be performed (from Civita, 1990a).

activities linked to land use and transformation are concerned. This, in turn, requires that the local and/or national authorities provide and maintain a service for updating these maps.

Until recently, the only possible approach to this type of mapping was a static scenario, a sort of situation picture at a given moment. Some efforts have been made to try "moving" the picture. Worth mentioning is the solution devised by B.R.G.M. (1976) for vulnerability mapping of France at 1:50 000 scale. For these maps, two transparencies showing time-dependent parameters related to human activities and groundwater exploitation are prepared and updated in time to overlay on a base, color-printed map of intrinsic vulnerability. However, only the base maps are available for sale; the transparencies are merely for reference purposes within the B.R.G.M.

Only with the recent advances of information systems has it been possible to build reliable real-time dynamic scenarios. Through this process, specific vulnerability maps can be regularly updated in a relatively short time. Although the newest systems radically change the classical mapping, they do not entirely change all the previous concepts. For instance, the idea of map scale must be converted into distribution and density of basic data. In fact, using AM/FM, CAD, and even more, GIS systems, scale becomes an operator- or/and user-selected option. The map itself is no longer a drawing but rather a data base. It becomes unnecessary, then, to print it at a given scale with a color code and given symbols unless it is needed for a specific goal.

**Map Construction Techniques**

Vulnerability maps are created manually or photographically, if they are on transparencies; or by computer, if the maps are encoded into any of several geographical information systems (GIS) such as ARC/INFO, ERDAS, or GENAMAP. The past few years have seen an increasing use of computers in the compilation of maps, which changes not only the nature of map production but also the very concept of the maps.

***Manual Techniques***

A vulnerability map may be manually drawn in various ways according to the survey scale, number and type of ground data, and type of data processing. The most widely used method is the overlaying of several base maps or/and transparencies manually or by a photographic process. This method helps select homogeneous groups and subgroups to which a range or a value of intrinsic vulnerability, selected in advance, is assigned (Albinet and Margat, 1970; B.R.G.M., 1973; 1975-1979; and 1976; Civita, 1990b; I.G.M.E., 1976; Olmer et al, 1978; Subirana Asturias and Casas Posnati, 1984; Vrána, 1968).

The stages and steps in the construction of a groundwater vulnerability map using the hydrogeological complex and setting method are as follows:

*Stage 1*

1) selection of lithostratigraphical, structural, and topographical information and outline of a base map according to homogeneous hydrogeological complexes and units;

2) construction of a map (or transparency) of the soil and overburden, with special attention paid to composition, thickness, and permeability parameters;
3) construction of a map (or transparency) of stream network density;
4) identification of similar homogeneous areas by overlaying the three above-mentioned maps.

*Stage 2*
For every homogeneous area:
5) construction of a map (or transparency) of the depth to water based on average water level data;
6) construction of a map (or transparency) of hydrogeological characteristics of the aquifer as detailed as available data allow;
7) construction of a recharge map;
8) identification of homogeneous settings as to intrinsic vulnerability by overlaying products # 5, 6, and 7 on the Stage 1 map. Reference should be made, as much as possible, to basic documentation for well-selected cases (Civita, 1990b).

*Stage 3*
For the entire area:
9) construction of a map (or transparency) of aquifer hydrodynamic characteristics and geometry (average potentiometric contour lines, flow directions, groundwater divides);
10) construction of a map (or transparency) of existing and potential contamination sources, existing and potential sites for contamination entries, objects in need of protection (as proposed in Appendix A);
11) construction of a specific vulnerability map by overlaying products # 9 and 10 on the Stage 2 map.

The method has an overall validity and generality specifically designed to cover large and geomorphologically complex land areas through operational or schematic mapping. But it lacks flexibility because it requires that every setting be assigned a mean value of a number of parameters (depth to water, permeability, net recharge, etc.), although the parameters may significantly vary even in small areas.

***Computer Generated Mapping***
Systems like DRASTIC start with an identification of homogeneous settings by an overlay process, using point county system model (PSCM) (see Chapter 5). The DRASTIC system (Aller et al, 1987) subdivides an area into a regular square grid raster (15 m feet per side). The same system of value attribution to a discrete area (0.5 km per side) has been previously proposed by Villumsen et al (1983) who used a computerized rating system to construct the vulnerability map. Also within the SINTACS methodology (Civita, 1990a), the land area is divided into finite square elements (0.5 km per side) to which the rating value of single parameters and of three different weight strings are assigned.

A somewhat similar approach has been suggested by Haertlé (1983) in order to assign the weight to lithostratigraphical data from drilling. This technique, borrowed by several researches working on lowland area vulnerability (Civita, 1989; Civita, Chiappone et al, 1990), was the basis of a vulnerability map by Josopait and Schwerdtfeger (1979) in the Lower Saxony and Bremen territory in Germany.

***Using a Geographical Information System (GIS)***

The advances of computers have greatly enhanced the effectiveness of GIS for a wide variety of mapping, planning, and management needs. Computer-based GIS is a powerful tool for integrating and analyzing data obtained from a wide range of sources such as remote sensing, soil surveys, land surveys, water sampling stations, topographic maps, and the census data. In GIS systems, all types of geographically-referenced data are spatially registered so that multiple themes of data can be compared and analyzed together. Virtually any data that are, or can be, mapped (i.e., are geographically referenced) can be digitized and stored in the computer. Once stored, these data can be automatically validated, analyzed, extracted, reformatted, updated, and mapped in a format and at a scale designed to meet a specific need.

For many groundwater protection purposes a GIS may provide greatly increased efficiency in data handling, analytical capability, and display flexibility. The GIS based methods also are an increasingly common means to assess vulnerability of groundwater. Generally, a GIS procedure is utilized because of the greater flexibility and detail it offers. Use of GIS also facilitates updating of vulnerability maps as more data become available.

A GIS is an interactive system that may be used both in planning and in investigations. This computer system has a structure as follows:

- one or more input data acquisition and manipulation devices (keyboard, scanner, digitizer, file reader, etc.);
- an adequate electronic data processor (a minicomputer or more) to operate a complex of special software (DBMS - Data Base Managing System; TIN - Triangularized Irregular Network generation system; etc);
- a work station with high definition color screen for display and editing;
- one or more output drawing devices (plotter, printer).

The basic data on which vulnerability assessment and maps are based (see Chapter 5) can be introduced directly into a GIS in the form of:

- values for point variables (elevations, borehole records, water levels, depth to water, hydraulic conductivity, soil characteristics, etc);
- point features (wells, springs, monitoring stations, effluent points, chemical and waste storage sites, spill locations, etc.);
- continuous survey lines (geophysical profiles, hydrogeological cross sections);
- various maps (hydrogeological complexes maps, vegetation and soil maps, potentiometric surface maps, protection zone boundaries, etc.);
- lines expressing relationships or linear features (boundaries, structural features, groundwater divides, pipelines, roads, sewer lines, etc.);
- remote sensing data (continuous gray-tone or color images).

Numerical data expressed by contouring or derived from contour maps can be calculated and displayed in a contoured map format (topographical slope, depth to water, soil thickness, etc.). Information about selected factors, such as fracture and karst features density per unit area (fracture index and karst index), average precipitation, infiltration rates, infiltration index, or density of stream network, can be incorporated into the GIS. As mentioned previously, manual or computer-assisted analyses of remote sensing imaging can become components of GIS and compared with other information types and formats. Such an operation is conceptually defined in Mather (1991) as "...superposition of maps of various features, converted to a common scale and projected so as to allow the identification of regions that satisfy particular requirements".

Computer-assisted synthesis relies on a GIS being based on map data; each measurement or category of data is expressed as a value (or rating, weight, assessment code) at two-dimensional coordinates. Data can be put into vector format manually; semi-automatically, using a digitizing table; or automatically by a scanning system, incorporating a line-following software. Another GIS format is the grid, useful where maps contain thematic classes or where contoured data are available. The raster format is the best and the most widely used way to enter data and represent remote sensing images and color-coded thematic maps as small sized pixels.

Once in a computer-compatible format, it is possible to register all data sets as data layers with a common coordinate system and manipulate them to produce derivative maps and, finally, the intrinsic vulnerability map. Other layers of a data base are loaded (and, even more important, are steadily updated) with various point and non-point data that express potential impact of human activities on the environment (existing and potential contamination sources and entries, main objects needing protection). In this way, a specific vulnerability map can be compiled at any scale and in real time. The GIS simplifies compilation of various kinds of integrated vulnerability maps of various degrees of complexity (Civita, 1990a). For example, the basic intrinsic vulnerability map can be overlain by contamination source symbols of various sizes to produce a potential problem map (Zaporozec, 1985).

The number of vulnerability maps drawn by computerized methods (CAD or GIS) has been rapidly growing after a slow start in the late 1980s. Besides maps of ten U.S. counties chosen to demonstrate the DRASTIC method (Aller et al, 1987), GIS or combined GIS/DRASTIC maps produced in that period include for example, those by Aru et al (1990), Evans and Myers (1990), Lance et al (1990) (1990), Liddle et al 1989), Porcher (1989), and Whittemore et al (1987). The newer maps from the early 1990s include such as developed by Civita, Fisso et al (1992), Civita, Forti et al (1991), and Rundquist et al (1991). A GIS project is anticipated to integrate all data useful for vulnerability assessment of the Netherlands (Breeuwsma and van Duijvenboden, 1987).

However, considerations of cost, professional resources, and time may limit greater availability of vulnerability maps in a GIS format. The utility of a computerized GIS in vulnerability mapping depends on the program scope and constraints on time and budget.

If a local planning district or county needs to assess the effects of land-use planning changes on groundwater quality and if the study is not part of a long-term program or assessment, then a manual technique rather than a computerized, expensive, and time-consuming GIS may be appropriate. The high costs of digitization and manipulation of data, hardware, software, and personnel support may preclude the use of computerized GIS in some cases. A local study may be more efficiently conducted by the use of manual overlays of hydrogeological factor maps.

It is evident that both traditional and computer-based vulnerability maps will be needed for the foreseeable future.

## MAP DESIGN

The design of vulnerability maps still lacks international coordination and standardization. Maps, therefore, are not comparable on the global scale and their international understanding is at low level. Whatever construction method may be used (manual, computer-aided, automatic plotting), an agreement should be reached on colors, patterns, and symbols to be used, graphical design, and explanatory notes and text.

Based on the cumulative experience obtained from the vulnerability maps prepared in the past (see Chapter 4), an example of a layout of the map sheet is shown in Figure 19. This layout has been used for computer-aided (Civita, Forti et al, 1991), and CAD (Civita, Fisso et al, 1990) and GIS produced (Civita, Fisso et al, 1992) vulnerability maps.

The main vulnerability map is positioned in the top-center of the map sheet (Figure 19). Cross sections or block diagrams may be located below the main map. Some place must be available on the left to include special synoptical short legend. This legend has several columns (one for each degree of vulnerability). At the top of each column there is an explanation of the different degrees of vulnerability, increasing from right to left. Rectangular frames containing colors or/and color patterns representing selected vulnerability situations are put in the column corresponding to the appropriate vulnerability degree. A short description is included to the right of each frame.

At the bottom of the sheet one or more large-scale, supporting maps can be positioned to show specific information (e.g., existing contamination conditions of water bodies, groundwater resource quality and distribution, land use connected to diffuse agricultural contamination, stream network density, etc). If the vulnerability assessment used is a DRASTIC-type parametric system, small-scale maps of individual parameters may be positioned in this section (Schmidt, 1987; Civita, Forti, et al, 1991). Another solution is proposed by Zaporozec (1987): a section of the graphical design could contain numerical tables, each illustrating vulnerability assessment data of a selected setting using the DRASTIC system. Ferrara (1990) proposed pie charts depicting use and distribution of agrochemicals. Palmer (1988) added to this section explanation of a matrix system (see Figure 6) used to construct the vulnerability maps of the Severn-Trent Water Authority area.

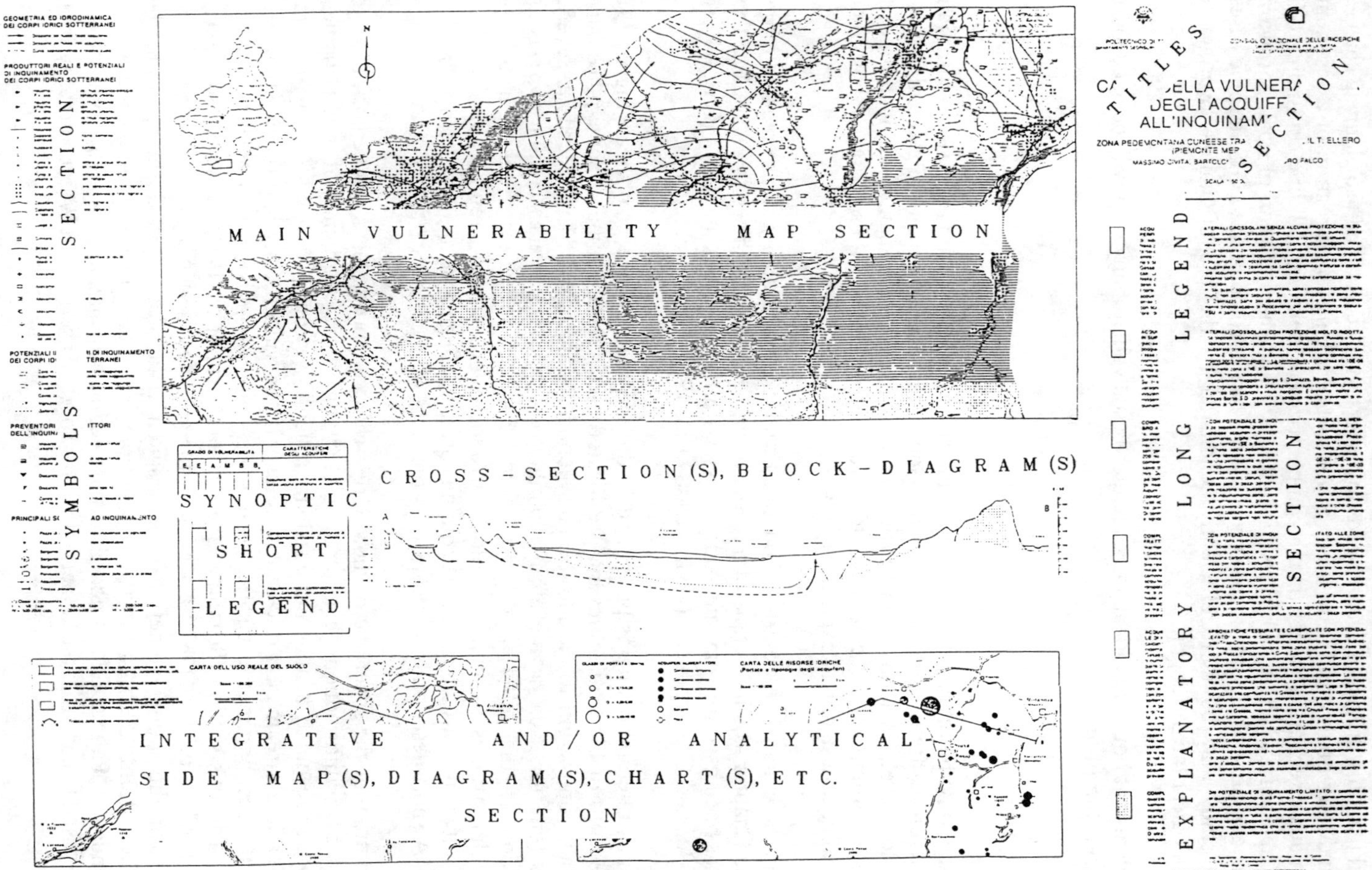


Figure 19. An example of a layout of a vulnerability map sheet (from Civita, Forti et al, 1991).

On the left side of the sheet, a legend containing symbols of hydrogeological features and human activities affecting the environment can be placed. The right-hand margin includes an extensive explanatory text with the same colors and patterns included in the synoptical short legend. The explanatory text describes hydrogeological complexes and settings and vulnerability features and conditions together with any other information useful for the map's design. This solution, adopted by several authors (Aureli et al, 1989; Civita, 1989; Martini and Marchetti, 1990; Schmidt, 1987; Zaporozec, 1987), combine both practical and economical criteria, excluding the editing of separate explanatory notes that would add excessively to the cost of printing. The top right-hand corner of the sheet contains map title, author's name(s), map scale, and the agency and/or authority that prepared the map.

Chapter 8.

# USES AND LIMITATIONS OF GROUNDWATER VULNERABILITY MAPS

## PURPOSE OF VULNERABILITY MAPS

Introduction of vulnerability maps in the early 1970s added a new dimension to presentation techniques of hydrogeological information. These maps express highly complex information in the form of a simple, intuitively understood term "vulnerability". As the name implies, the maps depict vulnerability, and therefore, skilled interpretation should not be needed to understand the message expressed by the map.

Groundwater vulnerability maps are valuable derivative maps that show, quantitatively or qualitatively, certain characteristics of the subsurface environment that determine vulnerability of groundwater to contamination. They are particularly useful for planning, regulatory, managerial, and decision-making purposes at all levels of government. Their primary purpose is to serve as guidelines for land-use zoning and the development of policy and strategy for groundwater protection and management. In fact, vulnerability assessment and maps constitute the first, essential step toward the protection of groundwater as a potential source of drinking water.

Vulnerability maps, when properly used, are valuable tools for environmental management. It should be stressed, however, that the vulnerability maps presently available should not be used beyond the limits specified in every single case. These maps should be viewed as one of the many tools for environmental management rather than as replacement of all other maps.

The fundamental concept of groundwater vulnerability is that some land areas easily contribute to groundwater contamination, and thus are more vulnerable, and others do not. Results of vulnerability assessment are portrayed on a map showing various homogeneous areas, sometimes called cells or polygons, which have different levels of vulnerability. The differentiation between the cells is, however, arbitrary because vulnerability maps only show relative vulnerability of certain areas to others, and do not represent absolute values.

When compiling a vulnerability map we should always keep on mind that vulnerability assessment and mapping will generally be used in conjunction with other related activities, such as water quality sampling, or analysis of risk to health or to the environment. In addition, from a policy perspective, decisions based on vulnerability mapping may result in an appeal process, and consequently, in court requirement for providing additional, more detailed or higher quality data. That may lead to reassessment of vulnerability.

Governmental officials can use the vulnerability map to aid in determining whether or where they should study potential groundwater problems more closely. The vulnerability map can be combined with land-use maps, groundwater quality data, and contamination source inventories to direct available financial and manpower resources to the most vulnerable areas. However, such maps and information are only supplemental tools in groundwater protection programs. The ultimate goal are the human impact control efforts concentrated on regulating land uses and on minimizing existing or potential contamination at the source.

## USES OF VULNERABILITY MAPS

Groundwater vulnerability maps are used for three main purposes: (a) planning, (b) contamination assessment, and (c) education.

### Planning

The main value of vulnerability maps is that they can be used as an effective preliminary tool for the planning, policy, and operational levels of the decision-making process concerning groundwater management and protection. First of all, vulnerability maps are valuable guides to planning and can help planners and regulators make informed, environmentally sound decisions regarding land use and protection of groundwater quality. Secondly, vulnerability maps can be used for the first-cut screening of an area for regional planning, which would allow planners to direct emphasis to areas of highest priority.

Vulnerability maps are a very important component in the prioritization of groundwater protection policy goals. They provide a method for local and state/country agencies and policy makers to set priorities for their protection efforts and for addressing groundwater problems, which would allow them to better distribute usually limited staff and funds to resolve these problems. Targeting intensive management planning efforts to the most vulnerable areas that pose the greatest risk to groundwater will maximize society's efforts to prevent the problems and to protect groundwater resources.

If the contaminating activities cannot be avoided, they should take place at locations where the potential for deterioration of the environment is smaller or can be tolerated. This pragmatic approach has been adopted in many countries and groundwater vulnerability mapping plays an important role in this process. All activities that represent a threat to the groundwater may be banned from certain areas either because the protective mechanisms at these locations are inadequate or because the value of groundwater at these locations is too high to take any risk.

### Contamination Assessment

Vulnerability maps are a good tool for groundwater professionals to make local and regional assessment of vulnerability potential, to identify areas susceptible to contamination, and to indicate the relative degree of concern and effort needed for more detailed assessment. Vulnerability maps help determine which areas may have groundwater problems and what types of site-specific data or studies are needed. Vulnerability maps also can be used for the design of monitoring networks and for the evaluation of contamination situations.

Large financial resources are spent on groundwater monitoring networks. The number of potential contaminants has significantly increased during the last few decades and it is not possible to monitor all the contaminants at all locations. Therefore, a proper design of a monitoring network and establishment of a monitoring program (techniques and frequencies of sampling, extent of measurements, etc.) are critical. Vulnerability maps can particularly help with respect to networks designed for monitoring groundwater quality threatened by human influences. Typically, the adverse effect of human stress will show first in the most vulnerable areas where the transport time from the surface to the aquifer is shortest (Kalinski et al, 1994).

Vulnerability maps are, in principle, helpful for the evaluation of nonpoint contamination cases due to the weak strength of the source and due the averaging effect of the large area involved. The best results are obtained if such maps are constructed for specific contaminants (e.g., diffuse contamination of groundwater by nitrate of agricultural origin; Palmer, 1988) and if the relations between parameters reflect the real world.

Applicability of vulnerability maps for the evaluation of point contamination situations is limited, mainly due do the large number of potential contaminants and due to the scale of vulnerability maps. Normally it would not even be possible to distinguish individual contaminated sites on these maps. Typically, the highly vulnerability areas are those where contamination transport is fast and attenuation is low. The general features of vulnerability maps can help to evaluate the potential impact of an accidental point contamination, caused for example, by a road accident involving truck transporting dangerous chemicals.

However, detailed analysis of point contamination cannot be performed using vulnerability maps. It is unrealistic to expect that ready-made vulnerability maps, covering large areas, and suitable for point contamination studies will be developed in the near future. None of the authors of the existing vulnerability maps recommends use of vulnerability maps for that purpose. They always point out that site-specific studies are indispensable.

**Education**

Vulnerability maps are useful for educating and informing planners, regulators, and decision-makers about groundwater protection and contamination prevention. Maps can also be used to educate the public and policy makers about aquifers being part of a larger, interconnected ecological system affected by human activities. For politicians and managers, maps showing vulnerability are of great value as a warning light in administrative cases when the risk of groundwater contamination is present. It is very important that the information about human impacts in the region is included on vulnerability maps together with the vulnerability classes.

Vulnerability maps create public awareness about environmental protection because the term "vulnerability" is very explicit and readily understood by the non-specialist. Most people will regard vulnerable groundwater reservoirs as worthy of protection from actual and potential threats. The process of creation of vulnerability maps is very educational and it teaches us about the complexity of environmental issues and about the limitations in our attempts to describe the real world.

### Technical Aspects of Uses of Vulnerability Maps

In order to have a broad spectrum of uses and applications, vulnerability maps should be consistent, comparable, standardized in a graphical and numerical expression, understandable, with a good legibility, and accompanied by descriptive legend and comprehensive explanatory notes, thereby helping overcome the gap that frequently exists between the scientific and lay communities. Vulnerability maps are not constructed for research purposes but first of all for practical uses. Therefore, they cannot be too sophisticated and overcrowded with data, which may lead to their misinterpretation or misuse, or, in case of low understandability, even to their nonuse.

The use and applicability of vulnerability maps are also influenced by cartographical methods, techniques, and processing. Users--mostly non-technical people with policy oriented background--are so far more familiar with traditional, manually produced maps. And, in the case of general vulnerability maps depicting natural (intrinsic) vulnerability, manually compiled maps, with three-dimensional diagrams and cross sections, will predominate in the near future. Also atlases and maps with numerous, superimposed transparent overlays will continue to be frequently used.

In the case of maps of specific vulnerability to contamination, users will increasingly demand and prefer maps portraying groundwater vulnerability for different spatial and temporal scenarios of contamination. These demands will be better met by computerized cartography, i.e. digitized two- or three-dimensional maps, grid maps, or block diagrams. This method of map production improves the usability and applicability of maps. It is less time consuming, flexible in the application of scale, and easily updated. It permits a combination of various scenarios depending on the user's request. Computerized mapping and data base-management-systems-integrated geographical information systems are being increasingly used for vulnerability maps in several European countries and in the United States of America.

## LIMITATIONS OF VULNERABILITY MAPS

The limitations of vulnerability maps are mainly caused by:

(a) Lack of representative data and their relation to the scale of the map.
(b) Inadequate description of the system.
(c) Lack of generally accepted methodology.
(d) Verification and control of vulnerability assessment.

## Available Data and Their Relation to Map Scale

The limitations of vulnerability maps generally are given by the purposes for which they were compiled and by their contents that control the scale. The biggest constraint is the amount and quality of data needed to construct a representative map. The amount of data is closely related to map scale. The overall utility of a vulnerability map is highly dependent on the scale at which the map has been compiled, the scale at which data were gathered, and the spatial resolution of mapping (National Research Council, 1993). The scale influences the accuracy of information, the level of generalization of data, and the value of the attributes and their parameters. The maps are only as good as the information and data upon which they are based and as the knowledge and experience of the map makers. However, a cost/benefit analysis may be helpful to determine the point of diminishing return at which the cost of data would exceed the value of information presented.

## Inadequate Description of the System

All kinds of hydrogeological maps have their limitations caused by our inability to accurately describe the complicated, heterogenous physical world, our description of which is typically based on the extrapolation from a restricted number of observation points. Hydrogeological maps depict the general trends, but as far as the detailed information is concerned, these maps are often inadequate. Vulnerability maps, which are based on the hydrogeological maps, will of course suffer from the limitations and shortcomings of these maps.

In construction of vulnerability maps the author's judgement plays an important role. If the principle of conservative assessment of vulnerability is adopted for every parameter utilized to define the vulnerability classes, than the final product will be too conservative and therefore of limited use.

## Lack of Generally Accepted Methodology

Many researchers agree on which parameters are relevant but they disagree on the methodology for combining these parameters into a vulnerability statement; neither terminology nor approach is standardized; given the same data base, different authors will arrive at different conclusions. Until now, it was not possible to develop a generally accepted methodology for the construction of vulnerability maps where all the relevant parameters and conditions are combined into a universal, objective, and generally accepted vulnerability class or category. The existing vulnerability maps are not fully comparable, due to map makers' subjectivity.

## Verification and Control of Vulnerability Assessment

The time scale for processes involved in groundwater vulnerability considerations is frequently so large, that we have limited chance of verifying our vulnerability assessment before it is too late. There are examples proving that it is possible to restore a seriously contaminated surface water (river or lake) to its original state. At the same time, only exceptionally, successful aquifer remediation has been reported.

Vulnerability assessments are being performed on both contaminated and non-contaminated aquifers. The information obtained from already contaminated aquifers may be used to calibrate and validate the vulnerability assessment procedures. However, such calibration and validation procedures would only be useful if the results could be utilized at as yet non-contaminated locations. Vulnerability indexing involves rather subjective, not physically based, calculations, and therefore, it is unlikely that these methods are valid under different conditions. In other words, fitting the weighting functions to match an existing contamination picture has limited value for improving our chances of predicting future developments at other locations due to the lack of physical meaning of the algorithms applied.

It is unfortunate that validation and verification of vulnerability maps can only be done after the damage to the aquifer has occurred. Taking into account that the damage could be long term, it is a high price to pay for a faulty management decision.

## MISUSES OF VULNERABILITY MAPS

The use of vulnerability maps is predetermined by their inherent deficiency--generalization of multifactor data. The amount of data and the map scale are in a delicate balance. Any attempt to disturb this balance, for example by a common mistake of enlarging the general map and presenting it as detailed information, would lead to gross errors. The major potential misuse of vulnerability maps is in attempting to extract site-specific information from or in applying site-specific problems to a map generated for regional planning.

Each type of vulnerability maps should only be used for the purpose for which it was produced. A site-specific, single-purpose vulnerability map may be constructed at the request of a single user who needs site-specific interpretations and decisions. A general map portraying the intrinsic vulnerability of principal aquifers is appropriate only for planning purposes at the regional or national level. Therefore, under no circumstances should the vulnerability maps be used as substitute for site-specific studies. They give only a first insight into vulnerability potential of an area, after which always a detailed, on-site study must be done.

The biggest difficulty with the use of vulnerability maps is explaining that despite technical limitations, there are many good applications and interpretations of them. The maps provide useful information on constraints and limitations of the environment that would assist regulators in proper management of groundwater resources. The maps are useful as long as the user understands their limitations and the criteria upon which they were developed; for example, the scale, assumptions used, implicit generalization, or lack of validation. Vulnerability maps should be carefully thought out and their meaning and degree of reliability fully explained. It is important that disclaimers appear on maps informing the user of the map limitations and of the intended use, and that a map is accompanied by sufficient documentation to fully describe the assumptions and methodologies used and the level of accuracy of presented information. With proper disclaimers, any vulnerability map can be used, even that one based on scanty data (Zaporozec, 1993).

An example of a disclaimer (Zaporozec, 1987):

> "This vulnerability map is designed for general and planning usage only. It shows the sensitivity of groundwater to contamination in a generalized way; local details have been generalized to fit the map scale. The map does not show areas that have been or will be contaminated, or areas that cannot be contaminated, and the map cannot be used for any site-specific purposes. Detailed studies of individual areas may be necessary when specific information is needed. Characteristics of individual contaminants or the likelihood of contaminant release have not been taken into account when constructing the map."

The greatest worry is the misuse of vulnerability maps by well-meaning but uninformed individuals or groups with little understanding of hydrogeology, groundwater sensitivity to human impacts, and vulnerability concept. Although concerns have been raised that vulnerability maps will not be interpreted correctly by non-technical persons, this group cannot be excluded from the group of map users. Every precaution has to be taken to guard against the potential misuse of maps by non-technical persons.

The best way to avoid the misuse of maps is through education of potential users, by involving users in map-making process, and by making sure that standard warnings and caveats are on all maps. Text explaining the limitations of maps, how to use and not misuse the maps, should accompany each map. An uniform and acceptable title, explanation, and description of a map can provide some degree of safeguard against blatant misuse. When the final product--vulnerability map--is easily understood, this will help ensure that it will really be used and correctly interpreted (Zaporozec, 1993).

Another way to minimize the potential for misuse is to periodically update the maps on the basis of new knowledge and data. Too often vulnerability maps are viewed as the "final word," when in fact, they are "living" documents. Without periodical updating, the degree of potential misuse and misinterpretation is much greater.

Although there is the concern about the possible misuse of vulnerability maps, the authors believe that the danger of misuse and misinterpretation is outweighed by the possibility that good or proper decisions would be made using the vulnerability maps. It is better to provide the best interpretation of the existing data that is possible with present capabilities than have done nothing out of fear of potential misuse.

Chapter 9.

# FUTURE TRENDS IN GROUNDWATER VULNERABILITY MAPPING

## HISTORICAL PERSPECTIVE

The early groundwater vulnerability maps classified land areas on an assessment of the degree to which the underlying groundwater was susceptible to human impacts. The vulnerability classes used were broad, relative, quantitatively imprecise, and subjective. Through the 1970s and 1980s, vulnerability maps of various kinds were produced, which ranged from those showing natural vulnerability of groundwater to those that included the known sources of contamination in the area. Such maps usually were general vulnerability maps in so far as they did not attempt to distinguish the degree of hazard posed by individual contaminants.

During the past twenty-five years the science of hydrogeology has evolved dramatically. Increasing concern for groundwater quality has meant that groundwater protection has become very important in many countries. Greater precision also has been introduced to the assessment of the quantitative aspects of groundwater systems. The widespread availability of computers now enables the easy and rapid handling of large amounts of data and the development of more realistic models. Moreover, the arrival of digital mapping techniques has revolutionized the speedy manipulation of data, which permits the rapid updating of existing maps as new information becomes available. These various changes make it possible to provide the means to develop vulnerability maps of greater sophistication and scientific precision and will lead to further progress in the near future.

## MAJOR ISSUES TO BE RESOLVED

Before considering possible future changes in vulnerability maps, there are a number of underlying issues that remain to be resolved. Three of the most important are:

1) Development of a generally recognized and accepted definition of vulnerability.

2) Agreement on a generally acceptable approach to vulnerability mapping and consistency in the use of methods and symbols expressing vulnerability on maps.

3) Testing the validity of vulnerability maps.

1) In developing the definition of vulnerability a number of authors stress that in addition to the intrinsic susceptibility of an aquifer to human impacts, there should be included the contaminant loading potential. This loading potential depends on the type of contaminant source; the mode of contaminant release; and the amount, rate, and type of contaminant.

2) Many currently available systems used in the construction of vulnerability maps rely heavily on traditional data on soil and rock characteristics and depth to bedrock and groundwater. However, they omit the dynamic aspects of vulnerability such as the magnitude and frequency of groundwater level fluctuations and the direction and velocity of groundwater flow.

Currently there are many methods being used for preparing groundwater vulnerability maps. Also the map scales and map symbols vary from country to country. It is extremely difficult to compare one map to another when they are based on noncompatible assessment and construction methods. The use of common sets of vulnerability maps would improve the consistency and comparability across similar studies. Such a consistent approach would be most desirable, but its implementation would encounter many technical, financial, and political problems, and would require a long period of time.

Mapping techniques would be relatively easy to coordinate if individual organizations would agree on standardized methods to obtain basic attributes (similar to chemical analyses standards) and on uniform approach to interpretation and assessment of their parameters. A standard set of attributes and map symbols also could be established with relative ease, and guidelines could be developed by an international committee. Model of a legend for groundwater vulnerability maps is in Appendix A.

A standardized scale would be most desirable because it would allow for compatibility of maps generated by various organizations. Even though it is possible to compare the maps of different scales by photographically enlarging or reducing the maps, the map contents usually do not allow this approach. Enlarging from the small or intermediate to the large scale will not provide sufficient and accurate detail. Reducing the large to the small scale would create a virtually illegible map. A formal agreement of individual agencies and organizations, at national and international level, would be required to achieve consistency in scale.

3) To date little has been done in testing the validity of vulnerability maps. Careful field monitoring will be needed to test predictions and thus enable further refinement of the assessment and mapping concepts.

## FUTURE TRENDS IN THE PRODUCTION AND USE OF VULNERABILITY MAPS

As groundwater vulnerability maps become more widely used, the need and challenge to improve them will result in experiments and changes. The new techniques introduced by the geographical information system (GIS) hold out exciting possibilities for the future of groundwater vulnerability maps and their use. At this stage the likely trends include:

- Improvement of vulnerability assessment methods and unification of symbols expressing groundwater vulnerability on maps.

- Use of greater quantitative precision in defining vulnerability classes based on increasing knowledge of contaminant transport.

- Improved modelling of groundwater systems with particular emphasis on better understanding of processes in the unsaturated zone.

- A move towards more specific vulnerability maps for individual contaminants or groups of contaminants.

- Increased emphasis on the production of large scale vulnerability maps, e.g. 1:10 000 and 1:25 000.

- Development of computer assisted mapping utilizing a GIS that will greatly improve the use of vulnerability maps in groundwater protection.

- Regular updating of vulnerability maps as new information becomes available by use of digital mapping techniques.

- The integration of vulnerability maps on a routine basis into local and regional planning procedures.

**ANTICIPATED PROBLEMS**

It is important to remember that a vital component for the successful use of a method is the availability of adequate basic data. It is reasonable then to foresee different countries passing through a series of stages in the production of groundwater vulnerability maps based on data availability. With limited data, only simple general maps may be possible to produce. These are useful provided they do not make extravagant claims. They can provide an initial screening of a region for planners. This enables the planners to eliminate the hydrogeologically most unsuitable ares for certain types of land use. General vulnerability maps may also be useful in the case of urban settlements, where many types of human activities are likely to be present. General vulnerability maps, however, may provide a false sense of security for developers in areas classified as less vulnerable when groundwater vulnerability classification is too simplified or it is not based on representative data.

However, the use of groundwater vulnerability maps for local and regional integrated land-use planning and for the protection of drinking water sources will undoubtedly be increasing in years to come. Especially maps of large scales (1:100 000 and less) have a good prospect of becoming an important document for governmental decision-makers provided the maps are easy to understand and based on solid and reliable data.

Standardization of general and specific vulnerability maps in methods of map construction and symbols will facilitate formulation of the requirements asked of the map makers by the map users. This feedback is particularly important for the specific vulnerability maps. At the same time, it is important to try to match the level of a problem with an available and appropriate level of answer. It is to be stressed that a simple problem may not require a sophisticated and often expensive remedy, at least in the first instance.

There is also the matter of time delay. With materials of persistent toxicity, there may be a build-up of contamination "time bombs" in areas of lower vulnerability. Whether an aquifer is contaminated in 50 days or 50 years, neither situation should be acceptable.

It would be tragic if a misinterpretation of vulnerability maps resulted in the contamination of the very groundwater the maps were produced to help protect. The pooling of experience gained from the use of vulnerability maps in many countries will provide a useful basis for ongoing change and improvement.

Appendix A.

# MODEL LEGEND FOR GROUNDWATER VULNERABILITY MAPS

## INTRODUCTION

The following legend is intended to facilitate the preparation of groundwater vulnerability maps in an internationally standardized form. The symbols and patterns given in this model legend are not to be considered as standards but as devices that are strongly recommended for use and, whilst every effort has been made to present symbols that cover all ordinary requirements, it cannot be considered to be all embracing.

The model legend has been prepared for the representation of groundwater vulnerability and is based upon the concept developed in the main text of this publication, to which the user should refer for further detail. Intrinsic vulnerability is defined in Chapter 2 as:

> "An intrinsic property of a groundwater system that depends on the sensitivity of that system to human and/or natural impacts."

Hence intrinsic vulnerability is dependent upon the following factors:

(a) The lithology and thickness[1] of the unsaturated zone.

(b) The lithology of the saturated zone.

(c) The nature of the soil zone[2].

(d) The potential for contaminant spreading within the aquifer under the pertaining flow conditions.

---

[1] Removal of significant quantities of the unsaturated zone by human activity or natural events will generally increase the vulnerability of the groundwater system.

[2] The soil zone is here deemed to consist of the loose weathered material composed of a mixture of varying proportions of organic matter and mineral particles, which covers much of the land surface of the Earth to a depth ranging from a few millimeters to a few meters and has a generally greater attenuation capacity for many potential contaminants than the underlying saturated zone.

## CARTOGRAPHICAL REPRESENTATION

### Primary Information

In order to develop a groundwater vulnerability map, it is useful to categorize the basic information relating to vulnerability into primary and secondary. The primary information relates to the intrinsic vulnerability of the groundwater system according to the overlying strata and is represented on a map by a full color shading (Table A1). As can be seen from the table, this aspect of vulnerability includes a consideration of the lithology (in particular, the vertical permeability) and thickness of the unsaturated zone.

In the case of diffuse contaminants a consideration of the nature of the soil zone can be important and, if the data are available, a soil classification system, indicating the leaching potential of soils, can be included by employing different tones of the colors used in Table A1 or by an ornament with differing intensities of shading. However, it should be recognized that the soil zone is normally relatively thin compared to the unsaturated and saturated zones and so this factor should not disproportionately influence the overall vulnerability of the system. The soil classification, when required, needs only be superimposed upon the extremely high, high, and medium classes of vulnerability given in Table A1. An example of a possible approach, based on that used by the National Rivers Authority of England and Wales (1992) is given in Table A1a; however, local data availability will often determine the approach adopted. For small scale maps (e.g. national maps at scales on the order of 1:1 000 000) and in areas where data are limited, the detailed classification given in Table A1a will be unsuitable and a simpler approach (e.g., use of the three main classes without subclasses) would be more appropriate.

The classification in Table A1a groups different soils into three classes based on soil physical properties that affect the downward movement of contaminants. These properties include: texture, structure, soil water regime, and the presence of distinctive layers such as raw peaty topsoil and rock or gravel at shallow depth. This classification can be applied across all aquifers within the extremely high, high, and medium vulnerability classes described in Table A1 by using either different intensities of the colors given in Table A1 or ornaments with different intensities of shading. The subclasses can be indicated by bounded regions with appropriate identifying letters within, i.e. H1, H2, H3, I1, I2, or L. Soil data availability and the scale of the final maps will determine whether the use of the subclasses described below is appropriate.

### Secondary Information

The secondary information relates to the potential for contaminant spreading within the groundwater system and is based on a consideration of the nature of the saturated zone. This information is superimposed as an ornament on the basic shading representing the primary information (Table A2). A special case is that of a non-aquifer, and the brown shading overrides any consideration of the unsaturated zone.

**Hydrogeological Features**

The representation of selected hydrogeological data will normally enhance the value of a groundwater vulnerability map. Table A3 contains a series of symbols for generally relevant data. Where possible the model legend is consistent with the UNESCO/IAH International Legend for Hydrogeological Maps (Unesco, 1983, revised edition). However, it has proved necessary to introduce some differences; e.g., in the vulnerability map model legend boreholes and springs are represented in blue, as opposed to the red of the International Legend for Hydrogeological Maps, in order that they might be visible when located on the highly vulnerable areas, which are shaded in red orange.

**Human Activity**

As implied in Chapter 2, there will generally be little value in a map of intrinsic vulnerability per se, and the concept of vulnerability needs to be displayed in "a fashion that makes it useful and convenient in the decision-making process" (Bachmat and Collin, 1987). In practice this requires the production of a specific vulnerability map, where the potential impact of land use and contamination sources is indicated. Thus a series of symbols are required to represent the impact of human activity (Tables A4 and A5). These symbols include "objects of protection", i.e. activities related to water supply/storage, and "potentially contaminating activities".

An immediate benefit following the production of a specific vulnerability map will be the indication of the presence of potentially contaminating activities within vulnerable areas, which would, in turn, indicate potential problem areas. This will assist in prioritization of areas for investigation and monitoring when funds are limited. Additionally, specific vulnerability maps can be used to assess planning applications for potentially contaminating activities that might threaten individual groundwater sources and/or the groundwater resource as a whole. These themes are discussed more fully in Chapter 8.

**Diagrams, Cross Sections, and Side Maps**

Chapter 7 refers to the presentation of vulnerability maps, including the use of cross sections, side maps, and a special synoptical legend. Such side maps and cross sections will generally include specific presentation of a restricted number of data sets (e.g., the current quality state of groundwater bodies, groundwater resource quality and distribution, land use related to diffuse agricultural contamination, streamflow density). In many cases suitable symbols for such maps are provided by existing international conventions (e.g., hydrogeological and hydrological) or the previous sections of this proposed model legend. However, for the existing quality of groundwater bodies the symbols given in Table A6 are recommended.

**Scale of Maps**

The proposed ornaments presented in the model legend are generally applicable to medium scale maps (e.g., between 1:25 000 and 1:200 000), although it is realized that particular circumstances may warrant exceeding these limits. For example, national "orientation maps", at scales on the order of 1:1 000 000 will essentially show intrinsic vulnerability with a limited number of ornaments to indicate major hydrogeological and human features; whilst maps at scales of 1:25 000 or larger, will be very specific in nature and possibly require further symbols than those provided here. For large-scale maps, the map-maker may wish to introduce different "categories" of individual objects to be represented by using different sizes of the same symbol. In order to avoid misunderstandings, not more than three sizes of the same ornament should be used; if more categories are needed, the symbol should be varied.

Groundwater vulnerability maps are complementary to, but should not be overprinted on, existing hydrogeological maps; these should be presented as independent sheets. However, in order to facilitate comparison, the map-maker may consider the production of vulnerability maps at the same scale as existing (or planned) hydrogeological and/or soil maps.

**Presentation**

To be of maximum value, the final legend employed must be clear, concise, and as complete as practically possible. The user must not be required to guess what is indicated by the colors and/or symbols used. It is strongly recommended that the map, legend, and explanatory notes should form an inseparable unit (i.e. be printed on one sheet). Brief explanatory notes can be placed in the map margin and, if the need arises, more detailed explanation can be given in a separate note or printed in sepia or grey on the reverse of the map.

Careful choice of colors is required in order to permit legibility of superimposed patterns and ornaments. Whilst bright colors are optically impressive, experience has shown that the use of less intense colors is generally more effective. It is common to find that the first printed draft of a map is too bright and adjusting of the colors is often necessary.

**Table A1.** Vulnerability of the aquifer system according to the overlying strata. (Principal information on map represented as full shaded color.)

| VULNERABILITY | COLOR | NATURE OF UNSATURATED ZONE STRATA | EXAMPLE |
|---|---|---|---|
| EXTREMELY HIGH | Red Orange | Ineffective and/or insignificantly thick or discontinuous | Fissured or highly karstic |
| HIGH | Rose | Highly permeable with unsaturated zone <2 m thick | |
| MEDIUM | Yellow | Moderate permeability ($k_v$=-$10^{3}$-$10^{-5}$); depth to saturated zone 2-20 m (or 2-50 m in karst with low karstic index) | Commonly unconsolidated formation |
| LOW | Light olive green | Low permeability; depth to saturated zone > 20 m | |
| VERY LOW | Dark olive green | Practically impermeable and of significant thickness | Clay or shale |

**Table A1a.** Soil classification for use in determining aquifer vulnerability to diffuse contaminants (National Rivers Authority, 1992).

| CLASS | 1. Soils of High Leaching Potential (H) | 2. Soils of Intermediate Leaching Potential (I) | 3. Soils of Low Leaching Potential (L) |
|---|---|---|---|
| | Soils with little ability to attenuate diffuse sources of contamination and in which non-adsorbed diffuse sources of contamination and liquid discharges will percolate rapidly. | Soils that have a moderate ability to attenuate diffuse contamination sources or soils in which it is possible that some non-adsorbed diffuse contamination sources and liquid discharges could penetrate the soil layer. | Soil in which contaminants are unlikely to penetrate the soil layer because water movement is largely horizontal or soils have a large ability to attenuate diffuse contaminants. Generally, these are soils with a high clay content. It must be recognized that runoff from these soils may contribute to groundwater recharge elsewhere in the catchment. |
| SUBCLASS | H1. Soils that readily transmit liquid discharges because they are either shallow or susceptible to rapid bypass flow directly to rock, gravel, or groundwater. | I1. Soils that can possibly transmit a wide range of contaminants. | None. |
| | H2. Deep, permeable, coarse-textured soils that readily transmit a wide range of contaminants because of their rapid drainage and low attenuation potential. | I2. Soils that can possibly transmit non- or weakly-adsorbed contaminants and liquid discharges but are unlikely to transmit adsorbed contaminants. | |
| | H3. Coarse-textured or moderately shallow soils that readily transmit non-adsorbed contaminants and liquid discharges but have some ability to attenuate adsorbed contaminants because of their large clay or organic matter contents. | | |

**Table A2.** Potential of contaminant spreading within the aquifer system. (Secondary information represented by ornament upon the vulnerability shading of Table A1; based on National Rivers Authority, 1992.)

| RELATIVE POTENTIAL FOR CONTAMINANT SPREADING WITHIN AQUIFER | NATURE OF AQUIFER AND ORNAMENT | |
|---|---|---|
| HIGH | Karstified | |
| | Coarse gravel sediments | |
| | High fracture index | |
| MEDIUM | Low karst index | |
| | Medium gravel sediment | |
| | Medium fracture index | |
| LOW | Fine grained sediment | |
| | Low fracture index | |
| NON-AQUIFER | | Brown shading |

**Table A3** Hydrogeological features

## GROUND WATER AND SPRINGS

| | |
|---|---|
| 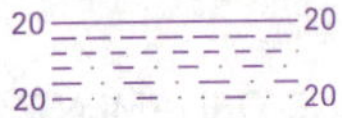 | Contours of the potentiometric surfaces (solid or broken lines with height relevant to reference level) |
|  | Direction of ground water flow |
|   | Ground water divide |
|  | Spring |
|  | Group of springs |
|  | Ground water seepage area |

## SURFACE WATER AND KARST HYDROGRAPHY

| | |
|---|---|
|  | Stream with perennial runoff |
|  | Stream with intermittent runoff |
|  | Dry valley, possibly with episodical runoff (ephemeral stream) |
|  | Karstic loss in river valley — no flow downstream |
|  | Aven |
|  | Doline filled with water |
|  | Dry doline |
|  | Lakes and reservoirs (irrespective of quality) |
|  | Site of ecological importance — e.g. wetland |
|  | River marsh |
|  | Bog |
|   | Scattered karstic forms (karst index = 0.5-1; 1-5; more than 5) |
| 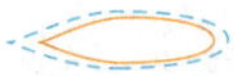 | Shott (playa) with episodical water |

**Table A4** Objects of protection

| | |
|---|---|
|  | Wellfield (L = multilayered aquifer system) |
|  | Well for potable water supply |
|  | Well for industrial or agricultural water supply |
|  | Ground water recharge site |
| 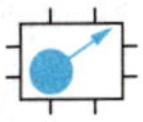 | Important spring developed for potable water supply |
|  | Important undeveloped spring |
|  | Thermal (T) or mineral (M) spring (or group of springs) |
| 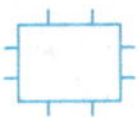 | Fenced perimeter of ground water development works |
|  | Limit of cone of depression resulting from ground water abstraction |
|  | Aqueduct |
|  | Drainage tunnel or trench for spring development |
|  | Underground storage for potable water |
|  | Source protection zone (pathogenic protection). The delineation of source protection zones will depend upon local practice and/or legislation |

**Table A5** Potentially contaminating activities

**MUNICIPAL**

 Urban area or large settlements, no sewerage network

 Urban area or large settlements, with sewerage network

 Main sewer trunk line

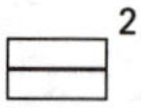

Treatment plant for urban/industrial wastewater (1 = primary, 2 = secondary, 3 = tertiary treatment)

 Collection point for non-treated urban or industrial sewage

Treatment plant for urban solid waste

Hospital

 Cemetery

**WASTE DISPOSAL**

 Cesspool, septic tank

Controlled landfill (letter indicates probable fill material: M = municipal solid waste; I = industrial or mining waste)

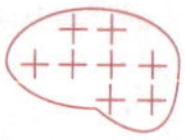 Uncontrolled and/or unauthorised landfill

Abandoned landfill (M = mixed solid, I = industrial, H = hazardous or toxic material, etc.)

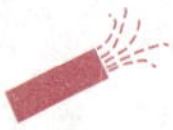 Spray irrigation of wastewater, whey, etc.

**Table A5** (*continued*)

**INDUSTRIAL**

| | |
|---|---|
| S | Industry with effluent of organic biological wastes (S = linked to urban sewerage) |
| S | Industry with effluent of marginally biodegradable wastes |
| S | Industry with effluent of inorganic wastes |
| | Oil/Fuel/Storage (garage/service station, mechanical workshop) |
| | Injection/disposal well |
| S | Chemical storage or stockpile (S = surface, U = underground) |
| G | Pipeline (G = gas, P = petroleum, C = chemicals, etc.) |
| | Thermoelectric power plant |
| | Nuclear power plant |
| | Hazardous or toxic chemical/waste spills, accidental or illegal |
| HTW | Treatment plant for hazardous and toxic wastes |
| S | Slaughterhouse |

**OTHER**

| | |
|---|---|
| | Highway, motorway, or railway |
| | Abandoned or improperly constructed well |
| | Airfield |
| | Military establishments |

**Table A5** (*continued*)

**MINING**

| | |
| --- | --- |
|  | Area of underground mining affecting the ground water regime |
|  | Area of open cast mining affecting the ground water regime |
| 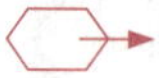 | Mine, pit (arrow indicates presence of a pumping plant) |
| 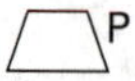 | Active quarry (P = excavation to piezometric surface) |
|  | Abandoned quarry |
|  | Filled quarry |

**AGRICULTURE**

| | |
| --- | --- |
|  | Animal husbandry with indication of number of units of manure |
|  | Livestock waste storage |
|  | Silage |
|  | Barren or untilled area (without use of pesticides, fertilizers, etc.) |
|  | Cultivated area with expected limited use of pesticides/fertilizers etc. |
|  | Cultivated area with expected frequent and abundant use of pesticides, fertilizers etc. |
|  | Flood irrigation area (e.g. rice-field, water meadow) |

**Table A6** Current quality state of ground water

| | |
| --- | --- |
| 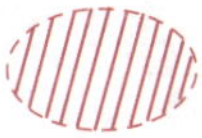 | Area with naturally poor ground water quality (requiring treatment for potable use) |
|  | Area with the natural quality of ground water altered by human activity |
| 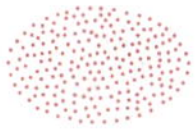 | Area with ground water contamination beyond national or international potable limits |
|  | Area with ground water contamination due to non-biodegradable organic compounds |
|  | Area with ground water contamination due to organic/biological matter |
|  | Area with ground water contamination due to inorganic compounds |
| 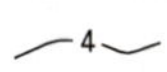  | Isoline defining contamination (units must be specified, e.g. mg/1, ppb, or °C) |

Appendix B.

# EXAMPLES OF GROUNDWATER VULNERABILITY ASSESSMENT AND MAPS

## EXAMPLES OF GROUNDWATER VULNERABILITY ASSESSMENT

The harmful influence of man on the environment cannot and will not be eliminated for many decades in the future. Many activities regarded as necessary to sustain our way of life, or even our existence, contribute to deterioration of the environment. It is obvious that this conflict cannot be solved by forbidding all contaminating activities. Therefore, it is necessary to develop attitudes and tools reducing the adverse effects of our activities on the environment.

It is our belief that vulnerability maps, when properly prepared and used, are valuable tools for environmental management. It should be stressed, however, that the vulnerability maps presently available should not be used beyond the limits specified in every single case. Although these maps should be included in the selection of tools for environmental management, they should not replace all other maps in the offices of environmental managers.

There are many different situations where groundwater is threatened. In principle, any stress exerted on the groundwater system can be taken into account in the assessment of vulnerability of an aquifer as it is shown in the five examples presented here. In spite of the fact that these examples cover a wide range of aquifer conditions and stresses, the vulnerability assessment procedure is the same in all cases.

Some of the examples show that, in spite of the many positive applications of ground water vulnerability assessment, there is a danger of faulty management decisions if these decisions are based solely on ready-made, generalized vulnerability maps.

## Example 1: Shallow Water Table Aquifer in Bangladesh

***Vulnerability of:***
Groundwater reservoir utilized by hand-operated, suction-mode pumps (groundwater is the only source of drinking water for the rural population of the region).

***Vulnerability to:***
Excessive pumping causing permanent or periodical lowering of the water table.

***System description:***
Under natural conditions the water table in a large part of Bangladesh is relatively high and groundwater is accessible to suction-mode pumps, which can lift water from 7-8 m below ground surface. As a part of the food self-sufficiency program, an increase of crop production requiring irrigation is being encouraged. As the result of the increased withdrawal of groundwater, the extent of the "deep-set" areas, where suction pumps cannot operate during dry seasons, becomes larger. An assessment of vulnerability is required to establish priorities with regard to providing a new type of pump being able to lift water from greater depth.

***Stress on the system:***
Combination of natural and human factors: deficit of precipitation during the period when it is most needed and an increased groundwater withdrawal for irrigation.

***Defense mechanisms of the system:***
In some cases high transmissivity of the aquifer will counterbalance the effect of irrigation water withdrawal by allowing groundwater from a larger area to contribute to the withdrawal. A good hydraulic contact between an aquifer and a river may be regarded as a defense mechanism reducing vulnerability of the aquifer. This contact would increase the vulnerability of a river, if the river was treated as a potentially vulnerable system.

***Key parameters and/or conditions:***
Hydraulic head distribution; infiltration rate; transmissivity; topography; and withdrawal rate.

***Secondary parameters:***
Water chemistry.

***Hydrogeological assessment of vulnerability:***
Vulnerability of groundwater reservoir to irrigation water withdrawal will vary significantly across the aquifer and will depend on the amount and distribution of the withdrawal, topography, and distance from the river. Parts of the aquifer situated far away from the river, corresponding normally to the area with higher ground elevations, and areas in the vicinity of wells should be regarded as more vulnerable.

***Vulnerability assessment using parametric methods:***
In principle, parametric methods are not applicable because no contaminant is concerned. If the assessment using parametric methods was made, the areas with the water table close to ground surface would be rated as highly vulnerable and the areas with the deeper water table as less vulnerable--a result entirely different from the results obtained by the hydrogeological assessment.

**Example 2: Unconfined Aquifer Exposed to Acid Rain, Denmark**

***Vulnerability of:***
Ground water reservoir.

***Vulnerability to:***
High load of acids in the infiltrating precipitation.

***System description:***
The aquifer is unconfined and characterized by a high carbonate content in the aquifer rock.

***Stress on the system:***
Due to utilization of coal with high sulfur content for energy production, there is a high content of sulfuric acid in the atmosphere. Rain contains, therefore, high concentrations of acid, which creates potential threat to groundwater quality.

***Defense mechanisms of the system:***
A part of the acid is neutralized during the passage through the soil and the remaining part is neutralized within the aquifer due to the presence of carbonate rock. The present rate of depletion of the carbonate is such that the acid will not cause decrease of the pH of groundwater within the next hundred thousand years.

***Key parameters and/or conditions:***
The distribution of the acid load in time and space; buffering and neutralizing properties of the soil and aquifer material; infiltration rate; and withdrawal distribution.

***Secondary parameters:***
Flow pattern in the aquifer (to determine dilution).

***Hydrogeological assessment of vulnerability:***
Vulnerability of groundwater reservoir toward acidification at the present stress situation is low.

***Vulnerability assessment using parametric methods:***
Carbonate content in the aquifer rock is the most important parameter in this situation, but most probably it would not be considered in preparation of the majority of vulnerability maps. The absence of carbonate would drastically increase vulnerability of this system.

***Remarks:***
Compare examples 2 and 3.

**Example 3: Unconfined Aquifer Exposed to Nitrates and Pesticides, Denmark**

***Vulnerability of:***
Groundwater reservoir.

***Vulnerability to:***
High load of nitrates and pesticides in the infiltrating water.

***System description:***
The aquifer is unconfined and characterized by a high carbonate content in the aquifer rock.

***Stress on the system:***
The area has a high level of agricultural activities and the rate of leaching of nitrate fertilizers and pesticides is high.

***Defense mechanisms of the system:***
The thickness of the protective soil layer is significantly reduced due to intensive agriculture and subsequent erosion. The soil organics are to large extent destroyed and the binding properties of the soil are small. Due to aerobic conditions in the unsaturated zone, no significant removal of nitrate takes place. Sorption and decomposition of pesticides in the soil zone and in the unsaturated zone is small. Removal of the nitrates and pesticides in the groundwater zone is minimal.

***Key parameters and/or conditions:***
Sorption capacity of the soil, unsaturated zone, and aquifer; infiltration rate; load estimates (nitrates and pesticides); flow pattern in the aquifer; and withdrawal distribution.

***Secondary parameters:***
Transmissivity; porosity.

***Hydrogeological assessment of vulnerability:***
Vulnerability of the aquifer in the present stress situation is very high for the whole area.

***Vulnerability assessment using parametric methods:***
The attenuation capability of the soil and unsaturated zone with regard to nitrates and pesticides is very small. The soil has been significantly removed by erosion, and the unsaturated zone delays, rather than reduces, the contaminant entry to groundwater. The parametric methods would erroneously assign lower vulnerability values to the parts of the aquifer with the thicker unsaturated zone.

***Remarks:***
Compare examples 2 and 3. The aquifer in this example is the same as the aquifer in example 2, but the stress on the system is different. The conclusion with regard to vulnerability rating of the aquifer is correspondingly different.

**Example 4: Hypothetical Alluvial Aquifer Exposed to Point Contamination**

***Vulnerability of:***
Hypothetical groundwater reservoir.

***Vulnerability to:***
Point contamination (accidently released toxic chemicals).

***System description:***
The aquifer (in a developing country) consists of alluvial deposits on two terraces: the lower terrace elevated 2-5 m above the level of the river and the higher terrace elevated 15-25 m above the river level. Drinking water supply for a local community is based on wells situated on the higher terrace. The present quality of ground and surface water is good. Groundwater discharges into the river at the concerned location. An industrialist, considering construction of a chemical plant at this location, has to obtain permit from local authorities. The governmental officials require that an assessment of vulnerability is made before the permit is granted because the plant represents potential risk for the village drinking water supply. The production requires a large amount of water and for economical reasons the investor is interested in placing the plant in the immediate vicinity of the river.

***Stress on the system:***
Contaminant flux caused by a hypothetical accident at the chemical plant.

***Defense mechanisms of the system:***
The aquifer consists almost entirely of a quartz sand, practically without attenuating properties for the contaminant considered. The soil is poorly developed and would not neutralize or adsorb any significant amount of the contaminant. Additionally, contamination would most likely originate below the soil level, and therefore, there is no geologically based protection of the aquifer.

***Key parameters and/or conditions:***
Infiltration rate; transmissivity; contaminant properties; and withdrawal pattern.

***Secondary parameters:***
Porosity; head distribution; river flow; and ground and river water quality.

***Hydrogeological assessment of vulnerability:***
Vulnerability would be high at all places where the contaminant would come into contact with the aquifer. Location of the plant on the higher terrace, far away from the river, will endanger the community water supply and will damage the aquifer for centuries. Location of the plant close to the river, in a groundwater discharge zone, will be an optimal solution for the aquifer protection. The damage to the aquifer would be confined to a relatively small area between the plant and the river.

***Vulnerability assessment using parametric methods:***
The thickness of the unsaturated zone on the lower terrace is small and the aquifer vulnerability is correspondingly high. The parametric methods would define the higher terrace as significantly less vulnerable than the lower terrace. Vulnerability assessment done by the parametric methods would place the plant at the wrong place.

**Example 5: Nubian Sandstone Aquifer, Arid Region**

***Vulnerability of:***
Groundwater reservoir.

***Vulnerability to:***
Drought and mismanagement.

***System description:***
The Nubian sandstone aquifer consists of 4 to 6 units separated by semipermeable layers. The aquifer sequence extends from Chad and Sudan (recharge areas) to Libya and Egypt (discharge areas). The part of the aquifer considered here is located under the Western Desert of Egypt. Several oases exist in the low-lying areas of the desert, where life depends on water supply from natural springs and/or deep wells drilled into different aquifer units. Precipitation in the area is negligible, around 5 mm/year, which normally evaporates before any infiltration to the aquifer can take place. In this area the hydraulic head typically increases with depth. Recently a significant lowering of head in several aquifer units has been observed and many wells dried out.

***Stress on the system:***
Increased groundwater withdrawal for domestic purposes in Libya and Egypt, and the excessive use of water in some of the oases caused lowering of the hydraulic head in the different units of the Nubian sandstone aquifer. The head losses are greatest at locations of artesian flow. An excessive irrigation combined with a high evapotranspiration rate resulted in creation of salt deposits at the ground surface damaging valuable land and posing potential threat to groundwater.

***Defense mechanisms of the system:***
There are no natural defense mechanisms protecting the aquifer system from drought and mismanagement.

***Key parameters and/or conditions:***
Groundwater flux across the aquifer boundaries at the Egyptian borders with Sudan and Libya; head distribution in the different aquifer units; transmissivity; storativity; infiltration distribution; and withdrawal pattern.

***Secondary parameters:***
Porosity; groundwater quality.

***Hydrogeological assessment of vulnerability:***
Vulnerability of the aquifer to drought is low, due to small amount (if any) of precipitation water that normally finds its way down to the aquifer. The decline in head is caused mainly by poor management, by water loss from flowing wells, and by excessive irrigation. Vulnerability of the aquifer system would be reduced if the water withdrawal took place from the higher aquifer units (where pumping is required) and if the higher aquifer units were replenished by water supplied by the lower aquifer units.

***Vulnerability assessment using parametric methods:***
Not applicable.

## EXAMPLE OF AN OPERATIONAL VULNERABILITY MAP

This example is a portion of the groundwater vulnerability map of East Kent (sheet 47), scale 1:100 000 (Figure B2), published by the National Rivers Authority in 1994. It is one of a series of 53 maps covering the whole of England and Wales that identify the vulnerability of groundwater to contamination.

To assess vulnerability, consideration has been given to the distribution of aquifers, to the physical and chemical properties of overlying soils, and to the characteristics of the unsaturated zone. An assessment of the physical and chemical properties of the soil is overlain, where appropriate, onto geological information, such as lithological type and permeability characteristics, to produce seven groundwater vulnerability classes. The map legend contains details of the soil and geological classifications and the way they are combined to give the vulnerability classification--these details are reproduced on the next page (Figure B1). For soil classification refer to Table A1a.

The vulnerability map series is a component of the National Rivers Authority's *Policy and Practice for the Protection of Groundwater* (1992) and will be important in the protection and management of aquifers. The approach and classifications used in the production of these vulnerability maps can also be used in the assessment of specific land-use practices, proposed developments, and land-use changes over aquifers where these could impact on groundwater quality.

## GEOLOGICAL CLASSIFICATION

Geological strata which contain groundwater in exploitable quantities are termed aquifers, whereas rocks which are largely impermeable and do not readily transmit water are termed non-aquifers. Aquifers vary in their general and hydraulic properties and can be classified as fractured, fracture-intergranular and intergranular and it is these properties, particularly in upper unsaturated zones, which form the basis of groundwater vulnerability assessment.

All groundwaters are controlled waters but it is convenient to subdivide permeable strata into two types; highly permeable (Major Aquifers) and variably permeable (Minor Aquifers). The former having generally greater capacity to transmit contaminated recharge entering at their surface than the latter.

Major Aquifers (Highly permeable)

These are highly permeable formations usually with a known or probable presence of significant fracturing. They may be highly productive and able to support large abstractions for public supply and other purposes.

On this map they comprise the Chalk and the Folkestone and Hythe Beds.

Minor Aquifers (Variably permeable)

These can be fractured or potentially fractured rocks which do not have a high primary permeability, or other formations of variable permeability including unconsolidated deposits. Although these aquifers will seldom produce large quantities of water for abstraction, they are important both for local supplies and in supplying base flow to rivers. Major Aquifers may occur beneath Minor Aquifers and on this map, in these situations, the underlying Major Aquifer is only shown where the overlying Minor Aquifer is a drift deposit.

Minor Aquifers, on this map, comprise the following solid formations: the Lenham Beds, Bagshot Beds, Claygate Beds, Blackheath and Oldhaven Beds, Woolwich Beds, Thanet Beds, Sandgate Beds, sands and limestones in the Weald Clay, Tunbridge Wells Sand, sand in the Wadhurst Clay and Ashdown Beds. They also comprise the following drift deposits: head gravel, coombe deposits, sand in clay-with-flints, blown sand, alluvium, dry valley deposits, river terrace deposits, marine and esturine alluvium (sand) and storm beach deposits.

Non-Aquifers (Negligibly permeable)

Formations which are generally regarded as containing insignificant quantities of groundwater form a third group. However, groundwater flow through such rocks, although imperceptible, does take place and needs to be considered in assessing the risk associated with persistent pollutants. Some Non-Aquifers can yield water in sufficient quantities for domestic use. Major or Minor Aquifers may occur beneath Non-Aquifers.

On this map they comprise the London Clay, Gault, Atherfield Clay, Weald Clay, clay in the Tunbridge Wells Sand, Wadhurst Clay and clay in the Ashdown Beds.

Superficial drift deposits which overlie the solid geological strata can sometimes be substantial in thickness. They are often variable in composition changing from highly permeable outwash gravels to low permeability clays over short distances both laterally and vertically. The presence of low permeability drift deposits at the surface is identified by stipple shading whereas more permeable drift deposits are recognised as Minor Aquifers except where these overlie a Major Aquifer and they then assume the status of Major Aquifer. Only the uppermost drift deposits are classified on the map and there is no attempt to show the thickness of geological strata.

Beneath Romney Marsh, between Hythe and Rye, the distribution of permeable marine sands and impermeable marine clays has been considerably simplified. The underlying strata has been treated as if consisting entirely of permeable sands but in practice it is known that impermeable clay formations are also present.

## VULNERABILITY CLASSES

| Geological Classes | Soil Classes |
|---|---|
| Major Aquifer (Highly Permeable) | High (H) 1, 2, 3, U* |
| | Intermediate (I) 1, 2 |
| | Low |
| Minor Aquifer (Variably Permeable) | High (H) 1, 2, 3, U* |
| | Intermediate (I) 1, 2 |
| | Low |
| Non-Aquifer (Negligibly Permeable) | |

Low permeability, non-water bearing drift deposits occurring at the surface and overlying major and minor aquifers are head, clay-with-flints, brickearth, peat and marine and estuarine alluvium (clay).

* – Because soil information for urban areas is less reliable and based on fewer observations than in rural areas, the worst case is assumed and such land is classified as high leaching potential (HU) until proved otherwise.

All maps involve a compromise between the representation of natural complexity and ease of interpretation of the map. Such compromises place limitations on the resolution and precision of map information. In this case, the variety of soils, geological strata and potential contaminants that have to be covered is wide, and the classification used is, of necessity, generalized. Individual sites and circumstances will always require further and more detailed assessments to determine the specific impact on groundwater resources. The maps only represent conditions at the surface and therefore where the soil and/or underlying formations have been disturbed or removed, for example during mineral extraction, the vulnerability class may have been changed. Hence, where there is evidence of disturbance there will be a need to determine groundwater vulnerability using site-specific data.

Map prepared by Cartographic Department, Soil Survey and Land Research Centre, Cranfield University, Silsoe, Bedford, MK45 4DT



**Figure B1.** Geological classification and vulnerability classes taken from the National Rivers Authority's 1:100 000 groundwater vulnerability map series of England and Whales. For soil classification refer to Table A1a.

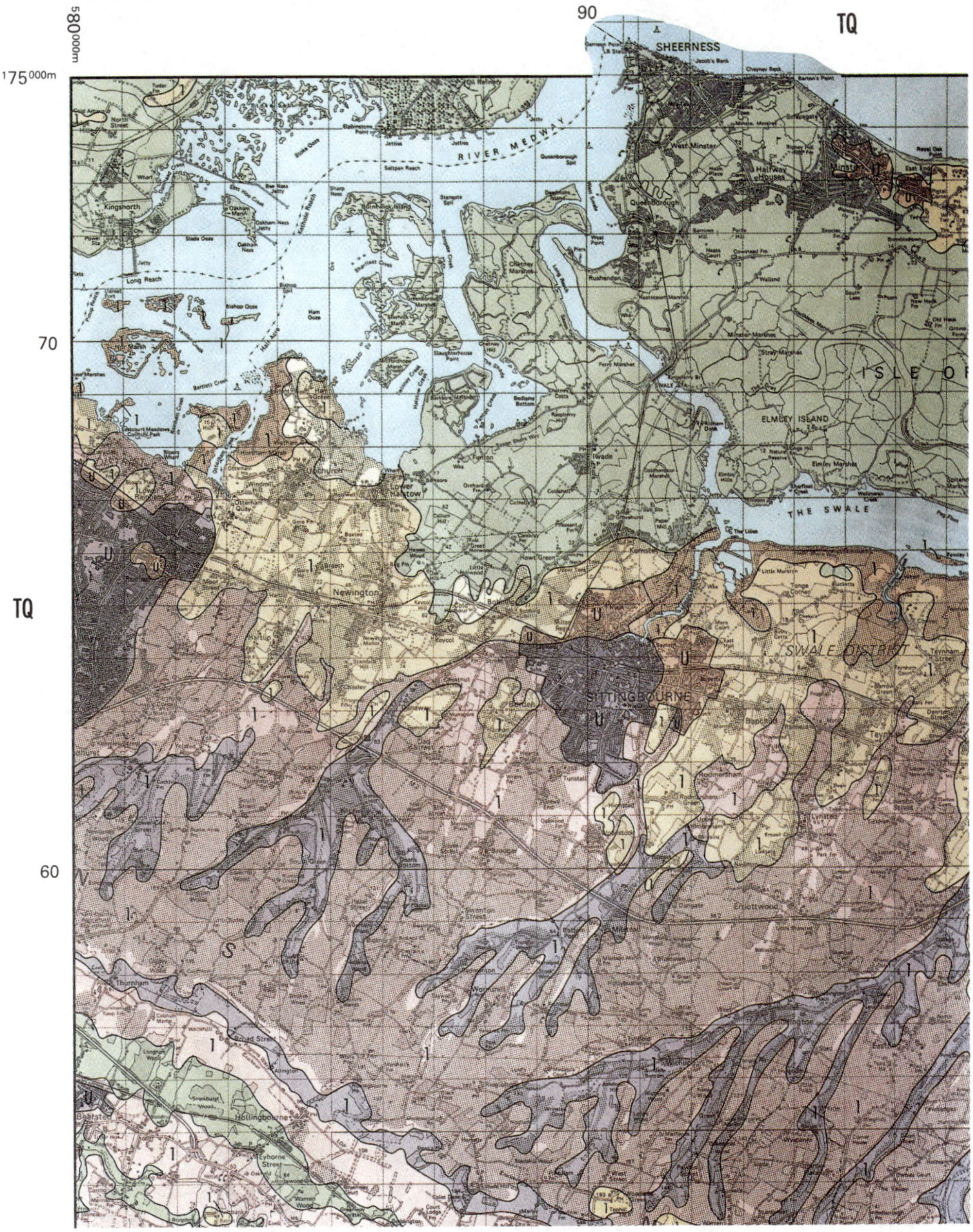


**Figure B2.** Part of the National Rivers Authority's groundwater vulnerability map of East Kent (sheet 47) at a scale of 1:100 000. The map shows the vulnerability of groundwater to contamination according to the key shown in Figure B1.

Appendix C.
# GLOSSARY

A conscious effort has been made to write in clear language, keeping technical jargon to a minimum. Even so, some terms are in such common use that their inclusion is considered to be justified. Problems also arise in two other areas. Some words are used with rather different meaning by individual authors and in different countries. For these words a particular meaning has been utilized in this publication as is explained in the glossary. There are also different forms of spelling of certain words, particularly between Britain and North America. Because the manuscript was largely prepared in the United States, American form of spelling was generally used.

The following technical terms were compiled from numerous sources and simplified to present their general meaning. More specific definitions are given in Bates and Jackson (1980), Fetter (1988), Pfannkuch (1990), Unesco (1991), and U.S. Geological Survey (1989).

| | |
|---|---|
| Absorption | Taking up, e.g. liquids in solids.<br><br>Compare: adsorption. |
| Adsorption | The attraction and adhesion of ions from an aqueous solution to the solid mineral surfaces with which it is in contact.<br><br>Compare: absorption, desorption.<br>Related term: ion exchange. |
| Advection | The process by which solutes are transported by the motion of flowing groundwater and at the same rate of flow. |
| Aerobic decay | Decomposition of organic substances, primarily by microorganisms, in the presence of free oxygen; the ultimate decay products are carbon dioxide and water.<br><br>Compare: anaerobic decay. |
| Anaerobic decay | Decomposition of organic substances in the absence or near absence of oxygen; the ultimate decay products are enriched in carbon.<br><br>Compare: aerobic decay. |
| Aquiclude | A hydrogeological unit that, although porous and capable of storing water, does not transmit it at rates sufficient to furnish an appreciable supply for a well or spring.<br><br>Obsolete term; see preferred term confining unit. |

| | |
|---|---|
| Aquifer | A rock unit that potentially yields groundwater to wells in exploitable quantities.<br><br>Compare: confining unit.<br>Related terms: confined aquifer, unconfined aquifer. |
| Aquifer sensitivity | The intrinsic susceptibility of an aquifer to contamination solely related to the hydrogeological characteristics of an aquifer and the overlying soil and geological materials.<br><br>Related terms: susceptibility, vulnerability. |
| Aquitard | A hydrogeological unit that retards but does not prevent the flow of water to and from an adjacent aquifer. It transmits water at a very slow rate compared to an aquifer.<br><br>Obsolete term; see preferred term confining unit. |
| Artesian well | A well in which the water level stands above the top of the aquifer but not necessarily above the land surface.<br><br>Compare: flowing well. |
| Attenuation capacity | The intrinsic ability of earth materials and an aquifer to absorb dilute, or retard contaminants. A complex of physical, chemical, and biological processes in the soil-rock-groundwater system (for processes see Table 1).<br><br>Related term: contamination potential. |
| Aven | A vertical shaft linking the land surface to a cave.<br><br>Partial synonym: vertical cave. |
| Catchment | An area that collects and drains rainwater.<br><br>Synonyms: drainage basin, watershed. |
| Confined aquifer | An aquifer bounded above and below by beds of distinctly lower permeability than that of the aquifer itself.<br><br>Synonym: artesian aquifer (obsolete term).<br>Compare: unconfined aquifer. |

**Confining bed (Confining unit)** A hydrogeologic unit of impermeable or distinctly less permeable material bounding one or more aquifers. A general term that replaces terms aquiclude, aquifuge, and aquitard.

Compare: aquifer.

**Contaminant plume** See plume.

**Contamination** Introduction into water of any undesirable substance not usually present in water, which renders the water unfit for its intended use. This change is not necessarily harmful to health.

Compare: pollution.
Related terms: contamination potential, plume.

**Contamination potential** Susceptibility of groundwater to contamination from a specific contamination source or by a specific contaminant.

Related terms: attenuation capacity, contamination.

**Desorption** Reverse process of adsorption.

Compare: adsorption.

**Diffusion** The process by which both ionic and molecular species dissolved in water move from areas of higher concentration to areas of lower concentration.

**Discharge** Flow of water from the aquifer.

Compare: recharge.
Related term: flux.

**Dispersion** The phenomenon by which a solute in flowing groundwater is mixed with uncontaminated water and becomes reduced in concentration. Dispersion is caused by differences in the velocity with which the water travels at the pore level and differences in the rate at which water travels through different strata in the flow path.

Related term: hydrodynamic dispersion.

**Doline** A circular depression in a karst area; commonly funnel-shaped, with wholly internal drainage.

**Effective porosity** — Ratio of the volume of interconnected pore space to the total volume of porous material.

Related term: porosity.

**Eh** — Redox potential (the energy gained in the transfer of 1 mol of electrones from an oxidant to $H_2$).

**Evapotranspiration** — A combination of evaporation and transpiration (evaporation of water by plants).

**Flowing well** — A well in which water level stands above the land surface and water flows without pumping.

Compare: artesian well.

**Flux** — The rate of flow (volume flow per unit area in unit time).

Related term: discharge.

**Freundlich isotherm** — An empirical equation that describes the amount of solute adsorbed onto a soil surface.

**Groundwater** — Subsurface water in the saturated zone below the water table.

**Groundwater divide** — Line on a water table or other potentiometric surface across which there is no groundwater flow and from which groundwater moves away in both directions.

**Groundwater flow** — Pattern of groundwater movement from recharge to system discharge.

Compare: groundwater system.

**Groundwater system** — An interconnected body of aquifers, usually of regional extent, which acts and can be studied as an unit.

Compare: groundwater flow system.

**Hydraulic conductivity** — A measure of the permeability of a rock.

Synonym: coefficient of permeability (obsolete term).
Related term: permeability.

**Hydraulic gradient** — The change in total head with a change in distance in a given direction. The direction is that which yields a maximum rate of decrease in head.

| | |
|---|---|
| Hydrological cycle | Perpetual movement of water in its different forms from the ocean through the atmosphere to the land and back to the ocean through various stages and processes.<br><br>Synonym: water cycle. |
| Hydrodynamic dispersion | The spreading of the solute front during transport through aquifer resulting from processes of mechanical dispersion and molecular diffusion. The spreading may be in the direction of flow (longitudinal) or perpendicular to it (transversal).<br><br>Related term: dispersion. |
| Infiltration | The flow of water downward from the land surface into the soil.<br><br>Compare: percolation. |
| Ion exchange | A process by which an ion in a mineral lattice is replaced by another ion that was present in an aqueous solution.<br><br>Related term: adsorption. |
| Karst | A type of topography that is formed on limestone, gypsum, and other rocks by dissolution, and that is characterized by sinkholes (dolines), caves, and underground drainage. |
| Percolation | The flow of water through earth materials.<br><br>Compare: infiltration. |
| Permeability | The ability of a rock or soil to transmit water.<br><br>Related term: hydraulic conductivity. |
| pH | The negative logarithm of the hydrogen-ion activity. |
| Piezometer | A small-diameter well installed to measure the elevation of the water table or potentiometric surface. |
| Piezometric surface | See potentiometric surface. |
| Plume | The spreading of a contaminant in the direction of groundwater flow from the point of origin to the point where contaminant concentration falls below the objectionable limits. The outer boundaries are in some cases difficult to detect. |

| | |
|---|---|
| Pollution | Introduction of objectionable material into water that may cause adverse health and environmental effects.<br><br>Compare: contamination. |
| Porosity | The void spaces in a rock; ratio of the volume of openings to the total volume of the rock.<br><br>Related term: effective porosity. |
| Potentiometric surface | The surface representing the water level in wells. The water table is the potentiometric surface of an unconfined aquifer.<br><br>Synonym: piezometric surface (no longer used; in older literature limited to the static level of water in a confined aquifer).<br>Compare: water table. |
| Precipitation | Water from atmosphere in the form of rain, snow, hail, or sleet. |
| Quaternary Period | The youngest period of the geologic time scale, forming together with the Tertiary Period the Cenozoic Era. It began two to three million years ago and extends to the present. |
| Recharge | The addition of water to the aquifer.<br><br>Compare: discharge. |
| Retardation | A general term for the many processes that act to remove the solutes in groundwater; for many solutes the solute front will travel more slowly than the rate of the advecting groundwater. |
| Saturated zone | The subsurface zone in which all interconnected openings are filled with water.<br><br>Synonym: zone of saturation.<br>Compare: unsaturated zone. |
| Shott | A shallow, intermittent, brackish or saline lake or marsh. |
| Soil | The upper 1 to 1.5 m of unconsolidated material. Contains living matter and supports plants. |
| Soil moisture | Water in the unsaturated zone. |
| Solute | The substance present in a solution; dissolved in water. |

| | |
|---|---|
| Spring | A discrete place where groundwater flows naturally from a rock or the soil onto the land surface or into a body of surface water. |
| Surface runoff | Water that flows over the land surface to the nearest stream or water body. |
| Susceptibility of groundwater to contamination | Lack of ability to resist the impact of contaminants on the quality of groundwater.<br><br>Compare: aquifer sensitivity, vulnerability. |
| Unconfined aquifer | An aquifer with the water table forming a free upper surface.<br><br>Synonym: water-table aquifer.<br>Compare: confined aquifer. |
| Unsaturated zone | The zone between the land surface and the water table that contains both water and air. It includes the soil water zone, intermediate zone, and capillary fringe.<br><br>Synonym: zone of aeration, vadose zone (obsolete term).<br>Compare: saturated zone. |
| Vulnerability | An intrinsic property of a groundwater system that depends on the sensitivity of that system to human and/or natural impacts. Two general types of vulnerability are differentiated: intrinsic vulnerability--depending solely on hydrogeological factors, and specific vulnerability--depending on hydrogeological factors and an imposed contaminant load.<br><br>Compare: aquifer sensitivity, susceptibility. |
| Wadi | The bed or channel of a stream in a desert region that is usually dry except during the raining season. |
| Water balance | Balance of input and output of water within a given defined hydrological area such as aquifer system, basin, lake, etc., taking into account net changes in storage.<br><br>Synonym: water budget. |
| Water table | The upper surface of the unconfined aquifer at which the pressure is about equal to the atmospheric pressure. |

Appendix D.
# REFERENCES CITED


Adams B. and Foster S.S.D. 1992. Land-surface zoning for groundwater protection. Jour. Institution of Water and Environmental Management, no. 6, p. 312-320.

Albertsen M. and Matthess G. 1978. Ground air measurement as a tool for mapping and evaluating organic ground-water pollution zones. Intl. Assoc. Hydrogeologists, Proc. Intl. Symp. on Ground Water Pollution by Oil Hydrocarbons, Praha, Czechoslovakia, p. 235-252.

Albinet M. 1970. Carte de la vulnérabilité à la pollution des nappes d'eau souterraine de la France à 1/1.000.000 [*Map of France of the ground water vulnerability to contamination, scale 1:1 000 000*]. Publ. DATAR-BRGM, Orléans, France.

Albinet M. and Margat J. 1970. Cartographie de la vulnérabilité à la polluttion des nappes d'eau souterraine [*Mapping of ground water vulnerability to contamination*]. Orléans, France, Bull. BRGM, 2ème série, section 3, n. 4, p. 13-22.

Aller L., Bennet T., Lehr J.H., Petty R.J., and Hackett G. 1987. DRASTIC: a standardized system for evaluating ground water pollution potential using hydrogeologic settings. U.S. Environmental Protection Agency, Ada, OK, EPA/600/2-87-036, 455 p.

Althaus H., Jung K.D., Matthess G., and Pekdeger A. 1982. Lebensdauer von Bakterien und Viren in Grundwasserleitern [*Lifetime of bacteria and viruses in ground water*]. Schmidt, Berlin, Umweltbundesamt, Materialien 1/82, 190 p.

Andersen L.J. and Gosk E. 1987. Applicability of vulnerability maps. In Vulnerability of soil and groundwater to pollutants (W. van Duijvenbooden and H.G. van Waegeningh, eds.), TNO Committee on Hydrological Research, The Hague, Proceedings and Information No. 38, p. 321-332; also printed in Environmental Geology and Water Sciences, 1989, 13/1, p. 39-43.

Antonov C. and Raikova B. 1978. Opit ea tipieirane na presnite pozemni vodi v nr Bulgarija po stepen na ujaesvimost. Chidrologija e Meteorologija, 27/6, p. 12-20.

Aru A., Barbieri G., Barrocu G., Chiarini E., Pani F., Sanna R.M., Uras G., and Vernier A. 1990. Applicazioni di cartografia automatica per la valutazione della vulnerabilita degli acquiferi di Oristano [*Application of automatic mapping to the assessment of aquifer vulnerability of the Oristano area*]. Proc. 1st Conv. Naz. "Protezione e Gestione delle Acque Sotterranee: Metodologie, Tecnologie e Obiettivi", Marano sul Panaro, v. 1, p. 41-60.

Aurand K., Matthess G., and Wolter R. 1971. Strontium-90, Ruthenium-106 und Caesium-137 in natürlichen Wässern [*Strontium-90, ruthenium-106 and caesium-137 in natural waters*]. Wiesbaden, Notizbl. hess L.-Amt Bodenforsch. 99, p. 313-333.

Aureli A., Adorni G., Chiavetta A.F., Fazio F., Fazzina S., and Messineo G. 1989. Carta della Vulnerabilità delle Falde Idriche - Settore Nord Orientale Ibleo (Sicilia S.E.) [*Map of aquifer vulnerability - Southeastern sector of the Hyblean area (SE Sicily)*]. S.E.L.C.A., Firenze., map 1:50 000.

Aust H., Vierhuff H., and Wagner W. 1980. Grundwasservorkommen in der Bundesrepublik Deutschland [*Ground water resources of the Federal Republic of Germany*]. Bonn, Serie "Raumordnung" des Bundesministers für Raumordnung, Bauwesen und Stadtebau, 06.043, p. 1-64, 3 maps 1:1 000 000.

Bachmat Y. and Collin M. 1987. Mapping to assess groundwater vulnerability to pollution. In Vulnerability of soil and groundwater to pollutants (W. van Duijvenbooden and H.G. van Waegeningh, eds.), TNO Committee on Hydrological Research, The Hague, Proceedings and Information No. 38, p. 297-307.

Bakour Y. and Kolars J. 1994. The Arab Mashrek, hydrologic history, problems and perspectives. Proc. Symp. on Water in the Arab World, 1-3 October, 1993, Harvard University Press, USA, p. 121-145.

Bates R.L. and Jackson J.A. (eds.). 1980 (2nd ed.). Glossary of geology. Falls Church, VA, Amer. Geological Inst., 751 p.

Beauduc P., Lienhardt M.J., and Rousselot D. 1976. Ground water vulnerability map 1:50 000, sheet Lyon. BRGM, Orléans, France.

Benacchio S. et al. 1988. Difesa degli acquiferi dell' alta pianura veneta, stato di inquinamento e vulnerabilità delle acque sotterranee del bacino del Brenta [*The protection of aquifers of the higher part of Venetian Plain, contamination assessment and ground water vulnerability of the Brenta River basin*]. Vol. 5°: Vulnerabilità dei corpi idrici soterranei all'inquinamento [*Vol.5: Vulnerability of aquifers to contamination*]. Monography G.N.D.C.I. - C.N.R. n. 207, Venice, 21 p., 2 maps, 1:50 000 and 1:25 000.

Breeuwsma A. and van Duijvenbooden W. 1987. Mapping the groundwater vulnerability to pollutants in The Netherlands. In Vulnerability of soil and groundwater to pollutants (W. van Duijvenbooden and H.G. van Waegeningh, eds.), TNO Committee on Hydrological Research, The Hague, Proceedings and Information No. 38, p. 309-320.

Breeuwsma A., Wosten J.H.M., Vleeshouwer J.J., Van Slobbe A.M., and Bouma J. 1986. Derivation of land quality to assess environmental problems from soil surveys. Jour. Soil Sci. Soc. of America 50, p. 186-190.

B.R.G.M. 1973. Carte hydrogeologique à 1/100.000 "Exploitation des reservoirs aquifères - risques de pollution", FF. Montpellier e Basse Vallée de l'Aude [*Hydrogeological map 1:100 000 "Utilization of aquifers - contamination risks", sheets Montpellier and Basse Vallée de l'Aude*]. B.R.G.M., Orléans, France.

B.R.G.M. 1975-1979. Cartes de vulnérabilité à la pollution des eaux souterraines (FF. 1/250.000 Lyon e Valence) [*Maps of ground water vulnerability to contamination 1:250 000, sheets Lyon and Valence*]. B.R.G.M., Orléans, France.

B.R.G.M. 1976... Cartes de vulnérabilité à la pollution des eaux souterraines (31 FF. 1/50.000) [*Maps of ground water vulnerability to contamination, 31 sheets, 1:50 000*]. B.R.G.M., Orléans, France.

Canter L.W., Knox R.C., and Fairchild D.M. 1987. Ground water quality protection. Lewis Publishers, Inc., Chelsea, MI, 562 p.

Carter A.D., Palmer R.C., and Monkhouse R.A. 1987. Mapping the vulnerability of groundwater to pollution from agricultural practice, particularly with respect to nitrate. In Vulnerability of soil and groundwater to pollutants (W. van Duijvenbooden and H.G. van Waegeningh, eds.), TNO Committee on Hydrological Research, The Hague, Proceedings and Information No. 38, p. 333-342.

Cavallin A., Cerutti P., Lauzi S., and Mancuso M. 1987. Valutazione quantitativa della vulnerabilità degli acquiferi mediante sistema automatizzato [*Quantitative assessment of aquifer vulnerability using an automatic system*]. Proc. Riun. Ricerc. Geol. Appl., Milano, p. 1-12.

Civita M. 1987a. La previsione e la prevenzione del rischio d' inquinamento delle acque sotterranee a livello regionale mediante le carte di vulnerabilità [*Forecasting and preventing the ground water contamination hazard on a regional scale using vulnerability maps*]. Proc. Congr. "Inquinamento delle acque Sotterranee: Previsione e Prevenzione", Mantova, p. 1-27.

Civita M. 1987b. L'inquinamento delle risorse idriche sotterranee in Italia: scenari, problematiche e strategie d'intervento [*The contamination of ground water resources in Italy: scenarios, problems and reclamation strategy*]. Proc. 1st Congr. Intern. di Geoidrologia "L'Antropizzazione e la Degradazione dell'Ambiente fisico", Firenze, 2-6 Dec., 1987, Publ. G.N.D.C.I. - C.N.R. n. 129, p. 299-306.

Civita M. (ed.). 1989. G.N.D.C.I. - C.N.R. Linea di Ricerca n. 4 "Valutazione della Vulnerabilità degli Acquiferi" [*G.N.D.C.I. - C.N.R. Research Line #4 "Aquifer Vulnerability Assessment"*]. Geda, Torino Report 1986/1989, 145 p.

Civita M. 1990a. La valutazione della vulnerabilità degli acquiferi all'inquinamento [*Assessment of aquifer vulnerability to contamination*]. Proc. 1st Conv. Naz. "Protezione e Gestione delle Acque Sotterranee: Metodologie, Techologie e Obiettivi", Marano sul Panaro, v. 3, p. 39-86.

Civita M. 1990b. Legenda unificata per le carte della vulnerabilità all'inquinamento dei corpi idrici sotterranei/Unified legend for the aquifer pollution vulnerability maps [in Italian and English]. G.N.D.C.I.-C.N.R. Pubbl. n. 276, Pitagora Editrice Bologna, 13 p.

Civita M. 1993. Ground water vulnerability maps: a review. Proc. IX Symp. on Pesticide Chemistry "Degradation and Mobility of Xenobiotics", Piacenza, Italy, October 11-13, 1993, Edizioni "Biagini", Lucca, Italy, p. 587-631.

Civita M., Chiappone A., Falco M., and Jarre P. 1990. Preparazione della Carta di Vulnerabilità per la rilocalizzazione di un impianto pozzi dell' Acquedotto di Torino [*Preparation of the vulnerability map to relocate a well field for the drinking water supply of Turin*]. Proc. 1st Conv. Naz. "Protezione e Gestione delle Acque Sotterranee: Metodologie, Tecnologie e Obiettivi", Marano sul Panaro, v. 2, p. 461-462.

Civita M., Cocozza T., Forti P., Perna G., and Turi B. 1983. Idogeologia del bacino minerario dell'Iglesiente (Sardegna Sud Occidentale) [*Hydrogeology of the Iglesiente mining district (SW Sardinia)*]. Mem. Inst. It. Speleologia, 2, 2, 137 p.

Civita M., Fisso G., Governa M.E., and Rossanigo P. 1990. Schema Idrogeologico, qualità e vulnerabilità degli acquiferi della Pianura Vercellese [*Hydrogeological outline, quality, and vulnerability of aquifers of the Vercelli Plain*]. Mem. G.N.D.C.I. - C.N.R. n. 273, e Provincia di Vercelli, 31 p.

Civita, M., Fisso G., Governa M.E., and Rossanigo P. 1992. Carta della vulnerabilità degli acquiferi all'inquinamento, Pianura di Vercelli [*Map of aquifer vulnerability to contamination, Vercelli Plain*]. Sistema informat. territor. e ambient. Provincia di Vercelli, edizione via ARC/INFO, CSI Piemonte (5 maps 1:50 000).

Civita M., Forti P., Marini P., Micheli L., Piccini L., and Pranzini G. 1990. Carta della Vulnerabilità degli acquiferi carsici delle Alpi Apuane [*Vulnerability map of karstic aquifers of the Apuan Alps*]. Proc. 1st Conv. Naz. "Protezione e Gestione delle Acque Sotterranee: Metodologie, Tecnologie e Obiettivi", Marano sul Panaro, v. 2, p. 465-468.

Civita M., Forti P., Marini P., Meccheri M., Micheli L., Piccini L., and Pranzini G., 1991. Carta della Vulnerabilità all'inquinamento degli acquiferi delle Alpi Apuane [*Aquifer vulnerability to contamination map of the Apuan Alps*]. Mem. explic., Monography G.N.D.C.I. - C.N.R. n. 399, Firenze, 56 p., map, 1:25 000.

Civita M., Giuliano G., Pellegrini M., and Zavatti A. 1987. Examples of groundwater pollution and vulnerability in the Po Valley. In Vulnerability of soil and groundwater to pollutants (W. van Duijvenbooden and H.G. van Waegeningh, eds.), TNO Committee on Hydrological Research, Proceedings and Information No. 38, The Hague, p. 347-353.

Collin J.J. 1989. An attempt to evaluate the socio-economic impact of hydrogeological maps. Memoirs of the Intl. Symp. on Hydrogeological Maps as Tools for Economic and Social Development, Hannover, May 30-June 2, 1989, p. 319-331.

Custodio E. 1990. Aquifer vulnerability to salinization and contamination: benefit from using environmental chemical and isotopic tracers in Spain. IAEA, Vienna, 28 p.

Dunlap W.J., Wilson J.D., Piwoni M.D., and Enfield C.G. 1983. Transport and fate of organic pollutants in the subsurface - current perspectives. Proc. Symp. on Biofilm Processes in Ground Water Research, Stockholm (NFR), p. 5-8.

Edmunds W.M., Darling W.T., and Kinniburgh D.G. 1987. Solute profile techniques for the estimation of recharge in semi-arid and arid terrains. In Recharge estimation in semi-arid and arid zones using unsaturated zone solutes. Proc. NATO Symp. Antalya.

Engelen G.B. 1985. Vulnerability and restoration aspects of groundwater systems in unconsolidated terrains in The Netherlands. Intl. Assoc. Hydrogeologists, Memoires, 18th Congress, Cambridge, UK, Vol. XVIII, pt. 3, p. 64-69.

Enquist P. 1989. The regional hydrogeological mapping of Sweden. Memoirs of the Intl. Symp. on Hydrogeological Maps as Tools for Economic and Social Development, Hannover, May 30-June 2, 1989, p. 92-96.

Evans B. and Myers W. 1990. A GIS-based approach to evaluating regional groundwater pollution potential with DRASTIC. Jour. of Soil and Water Conservation 45/2, p. 242-245.

Fenge T. 1976. Geomorphic aspects of sanitary landfill site selection. Victoria physical environment and development, H.D. Foster (ed.), Victoria BC, Western Geogr. Ser. 12, Chap. 7, p. 241-286.

Ferrara V. 1990. Carta della vulnerabilità all'inquinamento dell'acquifero vulcanico dell'Etna [*Vulnerability to contamination map of the volcanic aquifer of Etna*]. S.E.L.C.A. , Firenze, map 1: 50 000.

Fetter C.W. 1988 (2nd ed.). Applied hydrogeology. Merril Publishing Company, Columbus, OH, p. 565-580.

Foster S.S.D. 1987. Fundamental concepts in aquifer vulnerability, pollution risk and protection strategy. In Vulnerability of soil and groundwater to pollutants (W. van Duijvenbooden and H.G. van Waegeningh, eds.), TNO Committee on Hydrological Research, The Hague, Proceedings and Information No. 38, p. 69-86.

Foster S. and Hirata R. 1988. Groundwater pollution risk assessment. Pan American Center for Sanitary Engineering and Environmental Sciences, Lima, Peru, 79 p.

Fragiadiakis A., Klein W., Korte F., Moza P.N., Schennert I., Vockel D., and Weiss U. 1979. Verhalten von Organohalogenverbindungen in Systemen Pflanzen-Boden [*Behavior of halogenated organic compounds in plant-soil systems*]. Forschungsber. Jül.-Spez. (Jülich, Kernforschungsanlage) 45, p. 144-155.

Francani V. and Civita M. (eds.). 1988. Proposta di una normativa per l'istituzione delle fasce di rispetto delle opere di captazione di acque sotterranee [*A proposal of technical norms for the delineation of protection areas of the ground water sources*]. Geo-Graph, Milano, 277 p.

Freitag U. 1989. Cartographic annotations to an atlas of a river basin - the case of the Chang Jiang. Memoirs of the Intl. Symp. on Hydrogeological Maps as Tools for Economic and Social Development, Hannover, May 30-June 2, 1989, p. 120-121.

Fried J.J. 1987. Groundwater resources in the European Community, 2nd phase. Vulnerability - Quality (Synthetical Report) (unpublished report).

Friesel P. 1987. Vulnerability of groundwater in relation to subsurface behaviour of organic pollutants. In Vulnerability of soil and groundwater to pollutants (W. van Duijvenbooden and H.G. van Waegeningh, eds.), TNO Committee on Hydrological Research, Proceedings and Information No. 38, The Hague, p. 729-740.

Garrels R.M. and Christ C.L. 1965. Solutions, minerals, and equilibria. Harper & Row, New York, 450 p.

George W.O. and Hastings W.W. 1951. Nitrate in the ground water of Texas. Transactions Amer. Geophysical Union 32, p. 450-456.

Gerba C.P. and Keswick B.H. 1981. Survival and transport of enteric bacteria and viruses in ground water. Elsevier, Amsterdam, Studies in Env. Sci. 17, p. 511-515.

Golwer A. 1983. Underground purification capacity. In Ground water in water resources planning, Proc. UNESCO Intl. Symp., Koblenz, Germany, Aug. 28-Sept. 3, 1983, UNESCO/IAH/IAHS, National Committee of the Federal Republic of Germany for the IHP, vol. II, p. 1063-1072.

Golwer A. and Schneider W. 1973. Belastung des Bodens und des unterirdischen Wassers durch Strassenverkehr [*Deterioration of soil and subsurface water by road traffic*]. München, Gas- und Wasserfach 114, p. 154-165.

Goossens M. and van Damme M. 1987. Vulnerability mapping in Flanders, Belgium. In Vulnerability of soil and groundwater to pollutants (W. van Duijvenbooden and H.G. van Waegeningh, eds.), TNO Committee on Hydrological Research, The Hague, Proceedings and Information No. 38, p. 355-360.

Haertlé T. 1983. Method of working and employment of EDP during the preparation of groundwater vulnerability maps. In Ground water in water resources planning, Proc. UNESCO Intl. Symp., Koblenz, Germany, Aug. 28-Sept. 3, 1983, UNESCO/IAH/IAHS, National Committee of the Federal Republic of Germany for the IHP, vol. II, p. 1073-1085.

Hafen U., Merkel B., and Ullsperger I. 1989. Computer aided design of hydrogeological maps. Memoirs of the Intl. Symp. on Hydrogeological Maps as Tools for Economic and Social Development, Hannover, May 30-June 2, 1989, p. 567-575.

Handa B.K. 1983. Effect of fertilizer use on ground water quality in India. In Ground water in water resources planning, Proc. UNESCO Intl. Symp., Koblenz, Germany, Aug. 28-Sept. 3, 1983, UNESCO/IAH/IAHS, National Committee of Federal Republic of Germany for the IHP, vol. II, p. 1105-1119.

Holmberg M., Johnston J., and Maxe L. 1987. Assessing aquifer sensitivity to acid deposition. In Vulnerability of soil and groundwater to pollutants (W. van Duijvenbooden and H.G. van Waegeningh, eds.), TNO Committee on Hydrological Research, The Hague, Proceedings and Information No. 38, p. 373-380.

Ibrahim F.N. 1993. The ecological impact of groundwater use in the arid regions. Sahara and Sahel Observatory (SSO) Report, Paris, 44 p.

I.G.M.E. 1976. Mapa de la Vulnerabilidad a la contamination de los mantos acuiferos de la Espana Peninsular, Baleares y Canarias [*Aquifer vulnerability to contamination map of Spain Peninsula, Baleares, and Canary Islands*]. Primer Esquema cualitativo, 2 Edic. Mem. Explic., Serv. Publ. Min. Industria, Madrid, map 1:1 000 000.

I.G.M.E. 1979. Mapa de orientacion al vertido de residuos solidos urbanos, Zona de Almeria [*Orientation map for the selection of municipal solid waste sites, sheet Almeria*]. Serv. Publ. Min. Industria, Madrid, map 1:50 000.

Jackson R.E. (ed.). 1980. Aquifer contamination and protection. UNESCO, Paris, Studies and Reports in Hydrology 30, 439 p.

Johnson R.L., Keely J.F., Palmer C.D., Suflita J.M., and Fish W. 1989. Transport and fate of contaminants in the subsurface. U.S. EPA, EPA/625/4-89/019, 148 p.

Johnston, R.H. 1988. Factors affecting ground-water quality. National Water Summary 1986--Hydrologic Events and Ground-Water Quality, U.S. Geological Survey Water-Supply Paper 2325, p. 71-86.

Josopait V. and Schwerdtfeger B. 1979. Geowissenschaftliche Karte des Naturraum-potentials von Niedersachsen und Bremen [*Geoscientific map of natural environment potential of Lower Saxony and Bremen*]. CC 3110 Bremerhaven-Grundwasser. Niedersachsischen Landesamt für Bodenforschung, Hannover, map 1:200 000.

Kalinski R.J., Kelly W.E., Bogardi I., Ehrman R.L., and Yamamoto P.D. 1994. Correlation between DRASTIC vulnerabilities and incidents of VOC contamination of municipal wells in Nebraska. Ground Water 32/1, p. 31-34.

Khouri J. 1989. Development in hydrological mapping of the Arab region: a contribution to hydrogeological mapping in arid zones. Memoirs of the Intl. Symp. on Hydrogeological Maps as Tools for Economic and Social Development, Hannover, May 30-June 2, 1989, p. 358-368.

Khouri J. 1993. Impacts of intensive development of deep groundwater in the great basins of the SSO and Arab regions. Sahara and Sahel Observatory (SSO) Publ., Paris, France, 101 p.

Klaučo, S. 1987. Motion of pollutants in the groundwater as factor of their vulnerability. In Vulnerability of soil and groundwater to pollutants (W. van Duijvenbooden and H.G. van Waegeningh, eds.), TNO Committee on Hydrological Research, The Hague, Proceedings and Information No. 38, p. 403-407.

Kleczkowski A.S., Myszka J., and Pyrich J. 1973. Projekt mapy stanu zagrozen i ochrony wod [*Map of ground water contamination hazard and protection*]. Technika Poszukiwan Geologicznych 12, p. 45-46.

Knackmuss H.J. and Reineke, W. 1979. Ursachen für den verlangsamten mikrobiellen Totalabbau von halogen-substituierten Kohlenwasserstoffen [*Causes for the retarded total microbial degradation of halogenated hydrocarbons*]. Forschungsber. Jül.-Spez. (Jülich, Kernforschungsanlage) 45, p. 198-210.

Kreimer R. 1970. Contaminación de aguas subterráneas [*Ground water contamination*]. Rev. de la Dir. Nacional de Geol. y Minería. Buenos Aires, vol. VI, no. 22, p. 63-80.

Lance E.V., Bitner M.J., and Graves T. 1990. Vulnerability of Bernalillo County groundwater resources. HMWS Task 1.4 Rept., ABQR 33/003, CH2M HILL, Albuquerque, NM (Chapters paginated separately).

Langmuir D. 1972. Controls on the amounts of pollutants in subsurface waters. Earth and Mineral Sci. 42/2, p. 9-16.

Lavie J. and Putallaz J. 1974. Carte de vulnérabilité de la nappe de la Crau [*Vulnerability map of the Crau aquifer*] (Bouches du Rhone), 1:50 000. BRGM, Orléans, France.

LeGrand H.E. 1964. System for evaluating the contamination potential of some waste sites. Jour. Amer. Water Works Assoc. 56/8, p. 959-974.

LeGrand H.E. 1983. A standardized system for evaluating waste-disposal sites. National Water Well Assoc., Dublin, OH, 49 p.

Lemaire B. and Martin P. 1973. Ground water contamination hazard map 1:100 000, sheet Montpellier. BRGM, Orléans, France.

Lewis W.J., Foster S.S.D., and Drasar B.S. 1982. The risk of ground water pollution by on-site sanitation in developing countries. WHO/IRWCD Report 01-82.

Lewis M.A. and Robins N.S. 1989. Hydrogeological maps and the British Geological Survey. Memoirs of the Intl. Symp. on Hydrogeological Maps as Tools for Economic and Social Development, Hannover, May 30-June 2, 1989, p. 97-106.

Liddle S.K., Pratt R.W., and Breidenbach C.M. 1989. Ground-water vulnerability to agricultural chemicals: Geographic Information System (GIS) pilot project. Research Triangle Institute, Research Triangle Park, NC, RTI/7848/002-02F, 18 p.

Llamas R., Back W., and Margat J. 1992. Groundwater use: equilibrium between social benefits and potential environmental costs. Applied Hydrogeology 2/92, p. 3-14.

Macioszczyk A. and Plochniewski A. 1979. Izmenenie kachestva podzemnych vod na vodozaborach Polshi i gidrogeologicheskie osnovy yego prognozirovaniya [*Ground-water quality changes in watersheds of Poland and hydrogeological principles of their forecasting*]. Intl. Assoc. of Hydrogeologists, Memoires, Vol. XV, Congress of Vilnius, USSR, 1. Reports, p. 284-287.

Marcolongo B. and Pretto L. 1987. Vulnerabilità degli acquiferi nella pianura a nord di Vicenza [*Aquifer vulnerability of the plain northwards of Vicenza*]. Pubbl. G.N.D.C.I. - C.N.R. n. 28, 13 p., map 1:25 000.

Margat J. 1968. Vulnérabilité des nappes d'eau souterraine à la pollution [*Ground water vulnerability to contamination*]. Bases de la cartographie, (Doc.) BRGM, 68 SGL 198 HYD, Orléans, France.

Margat, J. 1989. Tendances contemporaines de la cartographie des eaux souterraines en France: pluralisme et specialisation [*Modern trends in ground water mapping in France: general and specific*] . Memoires of the Intl. Symp. on Hydrogeological Maps as Tools for Economic and Social Development, Hannover, May 30-June 2, 1989, p. 49-56.

Margat J. 1991. La vulnérabilité des eaux souterraines aux pollutions, conception-estimation-cartographie [*Vulnerability of ground water to contamination, design-assessment-mapping*]. EEC Institut Européen de l'eau, Paris, France, 11 p.

Margat J. 1992. Problemes specifiques aux nappes souterraines transfrontieres [*Specific problems of transboundary aquifers*]. PNUE-CEFIGRE Atelier sur la "Gestion Environnementale des Bassins Internationaux", Sophia-Antipolis, 28-30 April, 1992, 5 p.

Margat J. and Suais-Parascandola M.F. 1987. Mapping the vulnerability of groundwater to pollution; some lessons from experience in France. In Vulnerability of soil and groundwater to pollutants (W. van Duijvenbooden and H.G. van Waegeningh, eds.), TNO Committee on Hydrological Research, Proceedings and Information No. 38, The Hague, p. 433-436.

Martini E. and Marchetti G. 1990. Carta di Vulnerabilità degli acquiferi all' inquinamento della Valle Umbra Nord [*Map of aquifer vulnerability to contamination of the northern part of Umbra Valley*]. Proc. 1st Conv. Naz. "Protezione e Gestione delle Acque Sotterranee: Metodologie, Tecnologie e Obiettivi", Marano sul Panaro, Memoria GNDCI-CNR n. 256 e Regione Umbria, Enclosure Vol., map 1:25 000.

Mather P.M. 1991. Computer application in geography. J. Wiley & Sons, New York, 365 p.

Matthess G. 1974. Heavy metals as trace constituents in natural and polluted ground water. Geologie en Mijnbouw 53, p. 145-155.

Matthess G. 1982. The properties of groundwater. John Wiley & Sons, New York, 406 p.

Matthess G. 1990. Die Beschaffenheit des Grundwassers [*The properties of ground water*]. Gebr. Borntraeger, Berlin-Stuttgart/New York, 498 p.

Matthess G., Bedbur E., Gundermann K.O., Loof M., and Peters D. 1991. Vergleichende Untersuchung zum Filtrationsverhalten von Bakterien und organischen Partikeln in Porengrundwasserleitern [*Comparative investigation of the behavior of bacteria and organic particles during filtration in porous aquifers*]. Stuttgart/New York, Zbl. Hygiene 191, p. 53-97 and 347-395.

Matthess G. and Pekdeger A. 1981. Concepts of a survival and transport model of pathogenic bacteria and viruses in ground water. Elsevier, Amsterdam, Studies in Env. Sci. 17, p. 427-438.

Matthess G., Pekdeger A., and Schröter J. 1985. Behaviour of contaminants in groundwater. UNESCO/IAH/IUGS, Intl. Contributions to Hydrogeology, vol. 6, Heise, Hannover, p. 1-86.

Matthess G., Pekdeger A., and Schröter J. 1988. Persistence and transport of bacteria and viruses in groundwater - a conceptual evaluation: Jour. Contaminant Hydrology 2, p. 171-188.

Meinardi C.R. 1982. Kwetsbaarheidskartering van het grondwater in Nederland, toegelicht aan het EG Project [*Ground water vulnerability mapping in The Netherlands, in conformity with the EC project*]. Rijksinstituut voor Drinkwatervoorziening, 28 p., map 1:500 000 (unpublished).

Miller J.C. 1975. Nitrate contamination of the water-table aquifer by septic tank systems in the coastal plain of Delaware. In Water pollution control in low density areas, W.J. Jewell and R. Swan (eds.), University Press of New England, Hanover, NH, p. 121-134.

Miller J.C., Hackenberry P.S., and DeLuca F.A. 1977. Ground water pollution problems in the southeastern United States. U.S. EPA, Ada, OK, EPA-600/3-77-012, 361 p.

Ministry of the Flemish Community. 1986. Vulnerability map of the groundwater for the Flemish Region. D/1986/3241/50, 10 p.

Moore J.S. 1989. SEEPPAGE: a system for early evaluation of the pollution potential of agricultural ground water environments. U.S. Department of Agriculture, Soil Conservation Service, Geology Technical Note 5 (revision 1), 23 p.

National Research Council (U.S.). 1993. Ground water vulnerability assessment--predicting relative contamination potential under conditions of uncertainty. Committee on Techniques for Assessing Ground Water Vulnerability, National Research Council, National Academy Press, Washington, DC, 204 p.

National Rivers Authority. 1992. Policy and practice for the protection of groundwater. HMSO, London, UK, 52 p.

Neumayr V. 1981. Verteilungs- und Transportmechanismen von chlorierten Kohlenwasserstoffen in der Umwelt [*Distribution and transport mechanisms of chlorinated hydrocarbons in the environment*]. Wasser-, Boden- und Lufthygiene (Reimer, Berlin), Berichte 3, p. 24-31.

Olmer M., Prokop J., Břízová-Zoubková M., Michlíček E., Řezáč B., and Vrána M. 1978. Map for protection of ground water - example. Intl. Assoc. Hydrogeologists, Intl. Symp. on Ground Water Pollution by Oil Hydrocarbons, Praha, Czechoslovakia, map 1: 200 000.

Olmer M. and Řezáč B. 1974. Methodical principles of maps for protection of groundwater in Bohemia and Moravia, scale 1:200 000. Intl. Assoc. Hydrogeologists, Memoires, Tome X, Congres de Montpellier, 1. Communications, p. 105-107.

Ostry R:C., Leech R.E.J., Cooper A.J., and Rannie E.H. 1987. Assessing the susceptibility of ground water supplies to non point-source agricultural contamination in southern Ontario. In Vulnerability of soil and groundwater to pollutants (W. van Duijvenbooden and H.G. van Waegeningh, eds.), TNO Committee on Hydrological Research, The Hague, Proceedings and Information No. 38, p. 437-445.

Palmer R.C. 1988. Groundwater vulnerability. Map 5 Lichfield. Soil Survey and Land Res. Centre, Cranfield, UK, 8 p., map 1:100 000.

Palmquist R.C. 1991. Ground water vulnerability: a DRASTIC approach. Washington Department of Ecology, Olympia, WA 98504, unpublished paper presented at the 84th Annual Meeting, Air & Waste Management Assoc., Vancouver, BC, Canada, June 16-21, 1991, 16 p.

Peczely T. 1976. Some observations concerning the change in the hydrological cycle caused by irrigation. In Proc. Symp. on Arid Lands Irrigation in Developing Countries, 16-21 February, Alexandria, Egypt, p. 63-88.

Pfannkuch H.O. 1989. Application of risk assessment to evaluate groundwater vulnerability to nonpoint and point contamination sources. Amer. Institute of Hydrology, Minneapolis, MN, Hydrological Science and Technology 5/1-4, p. 54-64.

Pfannkuch H.O. 1990. Elsevier's dictionary of environmental hydrogeology; in English, French, and German. Elsevier Science Publishers, Amsterdam, 332 p.

Pinneker E.V. 1974. The protection of the groundwaters of Siberia from pollution. Intl. Assoc. of Hydrogeologists, Memoires, Tome X, Congres de Montpelier, 1. Communications, p. 112-114.

Porcher E. 1989. Ground water contamination susceptibility in Minnesota. Minnesota Pollution Control Agency, St. Paul, MN, 29 p. and map 1:1 000 000.

Rassam J.C. 1988. Drought and low flows in the Arab states. WMO/UNPD/ACSAD Publ., Arab Center for Studies of Arid Zones (ACSAD) H.S./P/10, Damascus, Syria, 196 p.

Rogovskaya N.V. 1976. Karta yestestvennoi zashchischennosti podzemnykh vod ot zagrjaznenija [*Map of natural protection against contamination*]. Priroda 3, p. 57-61.

Rosen L. 1994. A study of the DRASTIC methodology with emphasis on Swedish conditions. Ground Water 32/2, p. 278-285.

Rundquist D.C., Rodekohr D.A., Peters A.J., Ehrman R.L., Di L., and Murray G. 1991. Groundwater vulnerability to contamination in Nebraska using DRASTIC method. Nebraska Conservation and Survey Division, Lincoln, NE, map 1:1 000 000.

Šarin A. 1989. What is a hydrogeological map and for what does it serve. Memoires of the Intl. Symp. on Hydrogeological Maps as Tools for Economic and Social Development, Hannover, May 30-June 2, 1989, p. 31-39.

Schmidt R.R. 1987. Groundwater contamination susceptibility in Wisconsin. Wis. Department of Natural Resources, Madison, WI, Wisconsin's Groundwater Management Plan Report No. 5, PUBL-WR-177-87, 27 p.

Schoeller H. (ed.) 1975. Ground-water pollution symposium. Intl. Assoc. Hydrol. Sci. Publ. 103, p. 1-7.

Schwille F. 1962. Nitrate im Grundwasser [*Nitrate in ground water*]. Koblenz, Deutsch. Gewasserkundl. Mitt. 6/2, p. 25-32.

Schwille F. 1976. Anthropogenically reduced ground waters. Hydrol. Sci. Bull. 21/4, p. 629-645.

Schwille F. 1981. Ground-water pollution in porous media by fluids immiscible with water. Elsevier, Amsterdam, Studies in Env. Sci. 17, p. 451-464.

Seller L.E. and Canter L.W. 1980. Summary of selected ground water impact assessment methods. Norman, OK, Natl. Center for Ground Water Research, Rept. 80-3, 142 p.

Sotorníková R. and Vrba J. 1987. Some remarks on the concept of vulnerability maps. In Vulnerability of soil and groundwater to pollutants (W. van Duijvenbooden and H.G. van Waegeningh, eds.), TNO Committee on Hydrological Research, The Hague, Proceedings and Information No. 38, p. 471-476.

Struckmeier W.F. 1989. Types and uses of hydrogeological maps. Memoirs of the Intl. Symp. on Hydrogeological Maps as Tools for Economic and Social Development, Hannover, May 30-June 2, 1989, p. 17-30.

Struckmeier W.F., Grimmelmann W.F., and Krampe K.D.W. (eds.) 1989. Memoirs of the Intl. Symp. on Hydrogeological Maps as Tools for Economic and Social Development, Hannover, May 30-June 2, 1989, 598 p.

Stumm W. and Morgan J.J. 1981 (2nd ed.). Aquatic chemistry. John Wiley & Sons, New York, 780 p.

Subirana Asturias J.M. and Casas Ponsati A. 1984. Mapa de vulnerabilidad a la contaminacion de los acuiferos del valle bajo del Llobregat (Barcelona) [*Map of aquifer vulnerability to contamination of the Llobregat valley (Barcelona)*]. Metodo de trabajo y estudio de la evolucion de las extracciones de aridos. Proc. 1st Congr. Espan. de Geol., 1, p. 795-809, map 1:10 000.

Tosan R., Lienhardt G., and Collin J.J. 1975. Ground water vulnerability map, sheet Lyon, 1:250 000. BRGM, Orléans, France.

Trojan M.D. and Perry J.A. 1988. Assessing hydrogeologic risk over large geographic areas. Univ. of Minnesota, St. Paul, MN, Minnesota Agricultural Experiment Station, Station Bulletin 585-1988, 65 p.

Unesco. 1970 (revised 1983). International legend for hydrogeological maps. UNESCO/IASH/IAH/IGS, 101 p.

Unesco. 1977. Hydrological maps. UNESCO, Paris and WMO, Lausanne, Studies and reports in hydrology 20, 204 p.

Unesco. 1991. International glossary of hydrology (draft - second edition). UNESCO, Paris, France, 561 p.

U.S. Environmental Protection Agency. 1987. Handbook: Ground water. U.S. EPA, Office of Research and Development, Robert S. Kerr Environmental Research Laboratory, Ada, OK, p. 12.

U.S. Environmental Protection Agency. 1990. Progress in ground-water protection and restoration. U.S. EPA, Office of Water, Washington, DC, EPA 440/6-90-001, p. 18.

U.S. Environmental Protection Agency. 1991. A review of methods for assessing the sensitivity of aquifers to pesticide contamination. U.S. EPA, Office of Ground-Water Protection, Washington, DC, preliminary document, p. 2-1.

U.S. General Accounting Office. 1991. Groundwater protection: measurement of relative vulnerability of pesticide contamination. U.S. General Accounting Office, Washington, DC, GAO/PEMD-92-8, 48 p.

U.S. Geological Survey, Subsurface-Water Glossary Working Group. 1989. Subsurface-water flow and solute transport, Federal glossary of selected terms. U.S. Geological Survey, Office of Water Data Coordination, Reston, VA, 38 p.

van Duijvenbooden W., Glasbergen P., and van Lelyveld, H. (eds.). 1981. Quality of groundwater. Elsevier, Amsterdam, Studies in Env. Sci. 17, 1128 p.

van Duijvenbooden W. and van Waegeningh H.G. (eds.). 1987. Vulnerability of soil and groundwater to pollutants. TNO Committee on Hydrological Research, The Hague, Proceedings and Information No. 38, 1143 p.

Van Schilfgaarde J. 1984. Drainage design of salinity control. Soil Salinity under Irrigation, Berlin, Springer Verlag, Ecological Studies 51, p. 190-197.

Vierhuff F. 1981. Classification of groundwater resources for regional planning with regard to their vulnerability to pollution. Elsevier, Amsterdam, Studies in Env. Sci. 17, p. 1101-1105.

Vierhuff H., Wagner W., and Aust H. 1981. Die Grundwasservorkommen in der Bundesrepublik Deutschland [*The ground water resources of the Federal Republic of Germany*]. Bundesanstalt für Geowissenschaften und Rohstoffe, Hannover, Germany, Geologisches Jahrbuch C30, 110 p. and 3 maps 1:1 000 000.

Villumsen A., Jacobsen O.S., and Sonderskov C. 1983. Mapping the vulnerability of groundwater reservoirs with regard to surface pollution. Geological Survey of Denmark, Yearbook 1982, Copenhagen, p. 17-38.

Villumsen A. and Sonderskov C. 1982. Vulnerability maps: a promising tool in groundwater protection. Aqua 5, p. 466-468.

Vladimirskij V.J. 1960. O zadachakh gidrogeologov v svyazi s kontrolem za ispolzovaniyem i okhranoi podzemnykh vod [*The role of hydrogeologists in ground water management and protection*]. Moscow, Razvedka i okhrana nedr 26/11, p. 34-37.

Vrána M. 1968. Ochrana prostých podzemních vod v Čechách a na Moravě, vysvětlivky k mapě 1:500 000 [*Ground water protection in Bohemia and Moravia, explanatory text to map 1:500 000*]. Water Resources and Plan. Center, Praha, Czechoslovakia.

Vrána M. 1984a. Methodology for construction of groundwater protection maps. (Lecture for UNESCO/UNEP Project PLCE-3/29, Moscow, Sept. 1981) published in Hydrogeological Principles of Groundwater Protection, E.A. Kozlovsky, Editor-in-Chief, UNESCO/UNEP, Moscow, vol. 1, p. 147-149.

Vrána M. 1984b. Novinky v mapování pro ochranu podzemní vody [*News in mapping for ground-water protection*]. Proc. VIIIth National Hydrogeological Conference "Puklinové a puklinovo-krasové vody a problémy ich ochrany" (V. Hanzel, ed.), Bratislava, Czechoslovakia, Geol. ústav D. Štúra, p. 379-383.

Vrba J. 1981. General aspects of ground water non-point pollution. Elsevier, Amsterdam, Studies in Env. Sci. 17, p. 41-53.

Vrba J. 1991. Mapping of groundwater vulnerability. Intl. Assoc. of Hydrogeologists, Ground Water Protection Commission, an unpublished working pater for meeting in Tampa, FL, USA, 35 p.

Wagner W. and Zomenis S.L. 1987. Vulnerability of major groundwater resources in Cyprus to contamination from the surface. In Vulnerability of soil and groundwater to pollutants (W. van Duijvenbooden and H.G. van Waegeningh, eds.), TNO Committee on Hydrological Research, The Hague, Proceedings and Information No. 38, p. 499-506.

Walker W.H. 1969. Illinois ground water pollution. Jour. Amer. Water Works Assoc. 61/1, p. 31-32.

Wang M. 1989. Hydrogeological mapping today. Memoirs of the Intl. Symp. on Hydrogeological Maps as Tools for Economic and Social Development, Hannover, May 30-June 2, 1989, p. 219-228.

Whittemore D.O., Merchant J.W., Whistler J., McElver C.D., and Woods J.J. 1987. Ground water protection planning using the ERDAS geographical information system: automation of DRASTIC and time-related capture zones. Proc. NWWA Focus Conference on Midwestern Ground Water Issues, Apr. 21-23, 1987, Indianapolis, National Water Well Assoc., Dublin, OH, p. 359-374.

Zampetti M. 1983. Informazioni e dati relativi alla quantità ed alla qualità delle acque sotterranee nella Comunità Europea [*Information and data on quantity and quality of ground water in the European Community*]. Proc. Intl. Conf. "Inquinamento delle Acque sotterranee da Composti organo-clorurati di Origine industriale", Milano, p. 197-204.

Zaporozec A. 1981. Ground-water pollution and its sources. GeoJournal 5.5, p. 457-471.

Zaporozec A. (ed.). 1985. Groundwater protection principles and alternatives for Rock County, Wisconsin. Wis. Geological and Natural History Survey, Madison, WI, Special Report 8, 73 p.

Zaporozec A. 1987. Groundwater pollution potential of Barron County, Wisconsin. Wis. Geological and Natural History Survey, Madison, WI, Map 87-2i, scale 1:100,000.

Zaporozec A. 1989. Hydrogeologic mapping for groundwater protection. In Recent advances in groundwater hydrology (J.E. Moore et al, eds.), Amer. Institute of Hydrology, Minneapolis, MN, p. 588-597.

Zaporozec A. 1993. Ground water vulnerability mapping practices in the United States. Wis. Geological and Natural History Survey (open-file report).

Zoeteman B.C.J., de Greef E., and Brinkmann F.J.J. 1981. Persistency of organic contaminants in groundwater, lessons from soil pollution incidents in The Netherlands. Elsevier, Amsterdam, Studies in Env. Sci. 17, p. 465-480.